Tropical Deforestation

Small Farmers and Land Clearing in the Ecuadorian Amazon

Tropical Deforestation

Small Farmers and Land Clearing in the Ecuadorian Amazon

THOMAS K. RUDEL
with Bruce Horowitz

COLUMBIA UNIVERSITY PRESS NEW YORK

Columbia University Press
New York Chichester, West Sussex

Library of Congress Cataloging-in-Publication Data

Rudel, Thomas K.

Tropical deforestation: small farmers and land clearing in the Ecuadorian Amazon/ by Thomas K. Rudel with Bruce Horowitz. p. cm.—(Methods and Cases in Conservation Science) Includes bibliographical references (p.) and index.

ISBN 0-231-08044-1(cloth)
ISBN 0-231-08045-x(paper)

1. Deforestation—Ecuador. 2. Deforestation—Amazon River Region. 3. Deforestation—Tropics. 4. Clearing of land—Ecuador. 5. Clearing of land—Amazon River Region. 6. Peasantry—Ecuador. 7. Peasantry—Amazon River Region. 8. Rain forest—Ecuador. 9. Rain forest—Amazon River Region. I. Horowitz, Bruce. II. Title. III. Series.
Sd418.3.E2R83 1993
333.75'137'0986—dc20 92-44356
CIP

∞

Casebound editions of Columbia University Press books are printed on permanent and durable acid-free paper.

Designed by Audrey Smith

Printed in the United States of America

c 10 9 8 7 6 5 4 3 2 1

To
Daniel and Esther Maria

Contents

Tables and Figures

Tables

Figures

Jacket illustration: The colonist village of San Carlos de Zamora shortly after the first settlers arrived (photo by M. Gildesgame)

Foreword

Mary C. Pearl

Tropical deforestation is a key concern in conservation because tropical forests are the most dense repositories of species, and because tropical forests supply both resources and ecological services to people. Too often, this concern has been expressed in the simplistic, urgent tones of direct mail advocacy: "As you read this sentence, 23 acres of forest have been destroyed." For those interested in protecting forests a statement of the global problem is insufficient. Rates of deforestation are widely variable across space and time. Forest removal or degradation is affected by myriad factors, including local people's numbers and cultures, the kinds of trees and other wildlife present, national policies, the weather, local markets, world markets, and the complex, idiosyncratic interactions of these and other factors. If we are to address effectively deforestation and its consequences, we must understand why it is happening. This volume explains why forest clearing has taken place in the Ecuadorian Amazon, taking into account the perspectives and actions of all the local actors over the past seventy years. In examining in such detail the ease of deforestation in one part of Ecuador, Rudel and Horowitz provide for us a framework for understanding deforestation in other parts of the tropics.

The series Methods and Cases in Conservation Science is designed to present to those with a professional interest in conservation examples of the most current and effective work in the field. The series emphasizes firsthand experience in order to ensure relevance to and inspiration for local conservation practitioners. Volumes in this series are designed to provide nonideological, specific information on issues or techniques that are key to effective conservation policy and practice. *Tropical Deforestation* fits squarely within that objective, and I hope it will elevate future analyses and subsequent management of forest use and forest destruction to a higher level of understanding.

Acknowledgments

I want to thank the large number of individuals and institutions who at various times during the past six years helped me with the research reported here. A fellowship from the Fulbright Commission in Ecuador provided much appreciated financial and logistical support in 1986. Small grants from the Rutgers International Food and Agriculture Program, the Rutgers Research Council, and the Rutgers Council on International Affairs financed the remote sensing analysis and follow-up trips to Ecuador. A Faculty Academic Study Leave from Rutgers University gave me the opportunity to convert some ideas and data into a book.

Bruce Horowitz provided absolutely indispensable assistance in collecting the data for this study. He carried out the field research in the Shuar village of Uunt Chiwias. My long association with colonists and my inability to speak Shuar made it impossible for me to do field research among the Shuar. The comparative focus of the book, on the Shuar as well as the colonists, stems directly from Bruce's work. He also made a large number of valuable comments on the first draft of the manuscript. Dr. Gonzalo Cartagenova and Ms. Maria Mogollon of the Fulbright Commission helped me get established in Ecuador when I returned to do field research in 1986. Padre Juan Bottasso gave me access to the archive of materials on the history of the Amazon region that the Salesian Mission in Ecuador has accumulated over the years. A series of officials at the Federacion de Centros Shuar gave us permission to carry out our study of land use in Uunt Chiwias. The staff of the Centro de Reconversion Economica de Azuay, Canar, y Morona Santiago allowed me to sift through and read their voluminous files on CREA sponsored settlements in the Upano—-Palora region.

Mike Gildesgame lent me his Peace Corps files on colonization and generously allowed me to reproduce his slides of Morona Santiago, some of which are printed in the book. Tom Yaccino graciously allowed me to look at his senior thesis on Sinai. Mike Bandiera wrote up his recollections of the early days of the Peace Corps—-CREA colonization program. Teuvo Airola, the Director of the Cook College Remote Sensing Center, introduced me to the mysteries of remote sensing and endured, along with Jim Gasprich and Shuang Chen, my unending questions about how to do this or that. Mike Siegel of the Rutgers Cartography Laboratory made all of the maps in the book and cheerfully put up with innumerable requests for small changes in the maps.

Lee Clarke, George Morren, and Claire MacAdams read and commented upon earlier versions of the argument presented in the book. Their critical eyes and efforts made this a better book. Emilio Moran read the entire manuscript and suggested some revisions that made a lot of sense. Harvey Molotch raised a set of questions about the theory, which made me think about deforestation in new ways. The anonymous reviewers at Columbia University Press suggested a number of useful revisions that I have incorporated. Baruch Boxer, Larry Brown, Christine Padoch, and Pete Vayda made valuable suggestions, which at different times pushed the book toward completion. Ed Lugenbeel, Amelie Hastie, and Laura Wood of Columbia University Press shepherded me through some revisions, edited the revised manuscript, and oversaw the production process with unfailing professionalism.

Susan Golbeck and Daniel Rudel helped me put issues of forest preservation in perspective with their constant reminders and invitations from life outside of the hermetic world of a writer. Thank you for keeping me sane! Finally, Bruce and I owe a huge debt to the hundreds of Ecuadorians from all races and stations in life who took the time to talk to us about forests, ecological preservation, and economic survival in a wonderful place on the eastern slope of the Andes. I wish we could help them in ways that are more tangible than the recording of their experience in a book.

Tropical Deforestation

Small Farmers and Land Clearing in the Ecuadorian Amazon

Introduction

■■

> There is . . . grandeur and solemnity in the tropical forest but little of beauty or brilliancy of colour. The huge buttress trees, the fissured trunks, the extraordinary air roots, the twisted and wrinkled climbers, and the elegant palms, are what strike the attention and fill the mind with admiration and surprise and awe. But all is gloomy and solemn, and one feels a relief on again seeing the blue sky, and feeling the scorching rays of the sun.
>
> —Alfred Wallace
> *A Narrative of Travels on the Amazon and Rio Negro (1853)*

Peasants felling trees in a rain forest cut poignant figures in the late twentieth century. Aspiring yeoman farmers, they struggle to better themselves by destroying an ecological treasure. The privations they experience begin with the search for land. Older peasants in Ecuador can recall when, in search of free land in the Amazon basin, they crossed the eastern range of the Andes on foot. To reach the forested lowlands, they endured days of wearying ascents and descents in and out of ravines, with ridges cutting jagged skylines above them. In one place where a trail emerges from the mountains, it crosses a five-acre patch of flat land that overlooks the jungle valleys below. The flat land so surprised the first migrants that they called the place Plan de Milagro, "plain of miracles." Settled in their new homes, the colonists faced additional problems: no income, deficient diets, and life-threatening illnesses. To commemorate these hardships, they named one new settlement Tristeza, "sadness."

The sights and sounds of the landscape change when colonists begin to clear land. When the trees fall, the skies open up, revealing ridges in the distance. The crack of axes and the buzz of chain saws resonate in the forest. Few sounds of birds and animals come from the remaining patches of the forest. Days begin with the blare of radios rather than the calls of toucans, parrots, monkeys, ocelots, and jaguars.[1] The place has changed, and something has been lost.

This book explores the difficult situation in which poor people do damage to the environment, and it attempts to explain their behavior. The emphasis on explanation addresses a persistent weakness in the research on tropical deforestation. Measuring the decline in the size of forests has proved easier than explaining why it has occurred. Life scientists have developed new, more accurate techniques for measuring deforestation, but they have not been able to agree about its causes. The disagreement stems in part from the bewildering variety of ways in which deforestation occurs. Fuelwood collection in alpine Nepal and land clearing by tractor in the Brazilian Amazon have very different origins. Even if we focus on deforestation in tropical moist forests, regional variations in the causal agents make it difficult to develop a single explanation. An expansion in cattle ranching explains much of the recent increase in deforestation in Latin America and none of the increases in deforestation elsewhere. Similarly, logging operations have played a major role in Southeast Asia and West Africa, but only a minor role in Latin America.

Some aspects of deforestation are easier to understand than others. The contribution of large cattle ranchers and logging companies to tropical deforestation seems clear (Browder 1988; Repetto and Gillis 1988; Hecht 1985). In contrast the contributions of peasants and small farmers to tropical deforestation remain mired in controversy. How much forest have peasants destroyed? Does population growth or proletarianization explain why peasants so often try to carve farms out of the rain forest? The confusion on these points is critical to our understanding of tropical deforestation because, by all accounts, peasants continue to clear large amounts of forest in Africa, Asia, and Latin America. This book attempts to untangle the confusion. Drawing on studies of deforestation on all three continents, it explains why and how peasants clear land. This explanation constitutes the core element in a theory of tropical deforestation. A case study of deforestation in the Ecuadorian Amazon illustrates the theory.

The book begins with a description of the changing geography of tropical deforestation. Its changing spatial coordinates suggest a theory that would account for the tremendous increases in deforestation during the twentieth century. Chapter 2 presents the theory. Chapters 3 through 7 illustrate the theory through a case study of deforestation in the Ecuadorian Amazon between 1920 and 1990. Chapter 3 describes the context for the case study with particular attention to government policies and the ecology of the rain forest; chapter 4 outlines how shifting coalitions of peasants, missionaries, and government officials overcame intermittent opposition from a well-organized indigenous group and deforested an extensive area in southeastern Ecuador; chapter 5 chronicles the struggles of one group of peasants to find and clear land in the northern part of this region; chapter 6 examines land clearing practices after settlement in two similarly situated communities, one inhabited by Shuar Indians and the other inhabited by mestizo colonists; and chapter 7 examines recent attempts by the communities' second generation to carve new farms out of the forest. Chapter 8 summarizes the theory, assesses its utility for explaining deforestation in other places, and spells out its implications for the design of policies to slow down tropical deforestation.

1

The Changing Geography of Tropical Deforestation

Because "deforestation" means different things in different places (Hamilton 1987:257), a description of the geography of tropical deforestation must begin with a definition which clarifies what does and does not constitute deforestation. In our usage tropical deforestation occurs when loggers clear more than 40 percent of the trees from a closed, primary forest or when small farms convert forests into fields or pastures.[1] When United Nations' officials first used this definition ten years ago (Lanly and Clements 1979), it generated some controversy. Critics charged that it did not take into account the tremendous loss in biodiversity which occurs when loggers 'high-grade' an area, extracting only 15 percent to 20 percent of the trees from an area (Jacobs 1988:11). This objection seems more convincing in theory than in practice. In Southeast Asia where loggers have done the most high grading, small farmers frequently follow the loggers into an area and begin to clear the degraded forests for agriculture. In these instances, high grading induces further land clearing, and these lands invariably become deforested by the U.N.'s definition.

Worldwide surveys of tropical deforestation report considerable variation from region to region in rates of deforestation. Latin America and Southeast Asia have experienced rapid deforestation while Africa, with the exception of several central and west African countries, has not (Lanly 1983). These regional differences in deforestation suggest a positive association between economic development and tropical deforestation. Other regional differences suggest that rates of deforestation vary with the size of the forests in a place. For example, in the western hemisphere deforestation has occurred more rapidly in Central America with its small, coastal forests than it has in the Amazon basin with its large blocks of forest.

Rain forests vary tremendously in their extent. In the Congo watershed, the Amazon basin, and the outer islands of Indonesia rain forests extend for hundreds and sometimes thousands of miles. In West Africa forests and cultivated lands form a mosaic with stretches of forest, usually at lower elevations, alternating with tracts of cleared land on plateaus. In Central America, Southeast Asia, and Central Africa small islands of forest ring the cones of volcanos in landscapes devoted to agriculture. Figure 1.1 outlines these differences in the shape and size of forests. The islands and blocks of forest in this figure represent extremes on a continuum of more to less fragmentation in forest cover.

The size and shape of forests influences how people clear land. Where small forests predominante, people clear land along the forest fringes. Where large forests prevail, people clear land along the fringes of the forest, but they also clear corridor-shaped tracts of land through the center of the forest. The people who clear the land and the events that spur them into action vary considerably from fringe to corridor clearing.

CLEARING ALONG THE FRINGES OF A FOREST

Panel 1 of figure 1.2 illustrates the two situations that give rise to this pattern. In one situation smallholders advance along a broad front

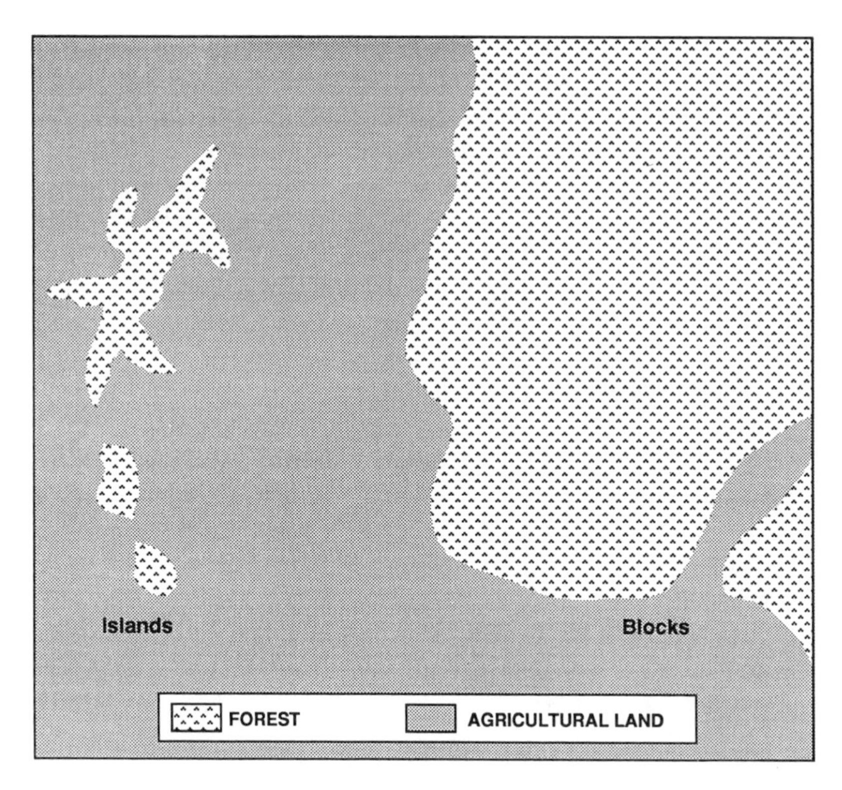

Figure 1.1. Variations in Forest Sizes and Shapes

into a forest. M. Dourojeanni provides a graphic description of how this type of deforestation proceeds in the upper reaches of the Amazon basin: "The population overflowing from the Andes down to the Amazon plains do not settle there. They advance like a slow burning fire, concentrating along a narrow margin between the land they are destroying and are about to leave behind, and the forests lying ahead of them" (1979, quoted in Myers 1984:150).

Agricultural expansion pushes the pioneer front into the forest. Keyes (1976) outlines the usual sequence of events in this process. Smallholders in northeastern Thailand prefer to clear uncultivated tracts of land closest to the village, but, when a village's population increases, household heads have to walk farther from the village in order to find uncultivated land. Eventually the distance between the village and the recently cleared fields becomes so great that the people cultivating these fields leave the original village and found a new village close to the fields. With continued population growth the cycle of agricultural expansion and village formation begins again with the new village as the point of departure. In another variant of this pattern, reportedly common in Latin America, settlers clear and farm a tract of land until yields begin to decline; then they cede or sell the land to cattle ranchers and move farther into the forest to begin again the cycle of forest clearing and abandonment (Stearman 1985:158).

In a second situation farmers clear land along the edges of small forests. Over the years these remaining islands of forest shrink in small increments as farmers expand the area under cultivation. Farmers walk farther from their homes to reach the forested lands, but the cultivation of the new land does not require that the families relocate. The slow eating away of the remaining islands of forest occurs in places with large, impoverished peasant populations, in Central America, the central African plateau, and the upland areas of Southeast Asia. Topography often limits the extent of the deforestation in these settings. The slopes on the side of a mountain become too steep to cultivate, or an impassable river makes farming in the upper reaches of a valley impossible.

CLEARING BLOCKS AND CORRIDORS OF LAND

Large landowners usually clear blocks of land (see panel 2 in figure 2.1). The deforestation that accompanied the creation of large coffee plantations in southern Brazil and banana plantations in Central America between 1900 and 1950 involved the conversion of large blocks of rain forest into fields. In many instances this type of deforestation occurs in vaguely defined corridors along a road or a river. For example, blocks of cleared land extend along both sides of the Cuiaba—Santarem road in Mato Grosso, Brazil (Lisansky 1990:54). In nineteenth-century Burma, plantations stretched up and down the rivers in the Irrawaddy delta, forming corridors of cleared land (Adas 1983).

The clearing of corridors of land in large forests has become common since 1960, but it has early origins. In the upper reaches of the Amazon basin the first spots of cleared land emerged in a linear pattern along mule trails from the Andes to the Amazon. Farther east navigable rivers provided access to markets, so the first clearings occurred in corridors of land along rivers. Similarly, access to markets via the sea contributed to the clearing of strips of land along the coasts of Central America and Southeast Asia. During the past thirty years land clearing in large forests has occurred primarily in corridors along recently constructed penetration roads (Hiroaka and Yamamoto 1980; Smith 1982). In recent satellite images of Rondonia in northwestern Brazil, deforestation occurs in cross-hatched corridors (see panel 3 in figure 1.2). The grid of roads in government sponsored colonization zones generates corridors of cleared land that run at right angles to one another. Islands of forest persist in the center of the squares created by the roads. In another pattern rows of farms appear along both sides of a road and at uniform distances back from the road. A second row of farms begins just behind the first row of farms, and the owners of the second row farms clear the land closest to the road on their farms. Their clearings form an additional corridor of cleared land that parallels the roadside corridor several kilometers into the forest (Hiroaka and Yamamoto 1980:431).

In another variant of this corridor pattern, common in the dipterocarp forests of Southeast Asia, loggers build a network of access roads in order to extract the commercially valuable timber. Clearing these cutover forests requires less work than clearing primary forests, and the new roads provide ready access to nearby markets, so numerous smallholders start farms along the roads (Vayda and Sahur 1985; Kartawinata and Vayda 1984). As the tropical wood trade has expanded during the past thirty years, this pattern of deforestation has spread over large areas of the Southeast Asian archipelago.

Every new road does not generate a corridor of cleared land in the rain forest. In peninsular Malaysia, an increasingly urban, affluent society, the recent construction of a road through a mountainous region did not generate additional pressures to clear land (Hurst 1990). In Colombia a penetration road into the province of Caqueta generated considerable amounts of land clearing while the construction of a similar road into the province of Guiviare did not stimulate much land clearing (Ortiz 1984:206). Similarly, road construction and paving in Rondonia, Brazil, had an enormous impact on forest clearing, while the construction of the Transamazon highway in Ama-

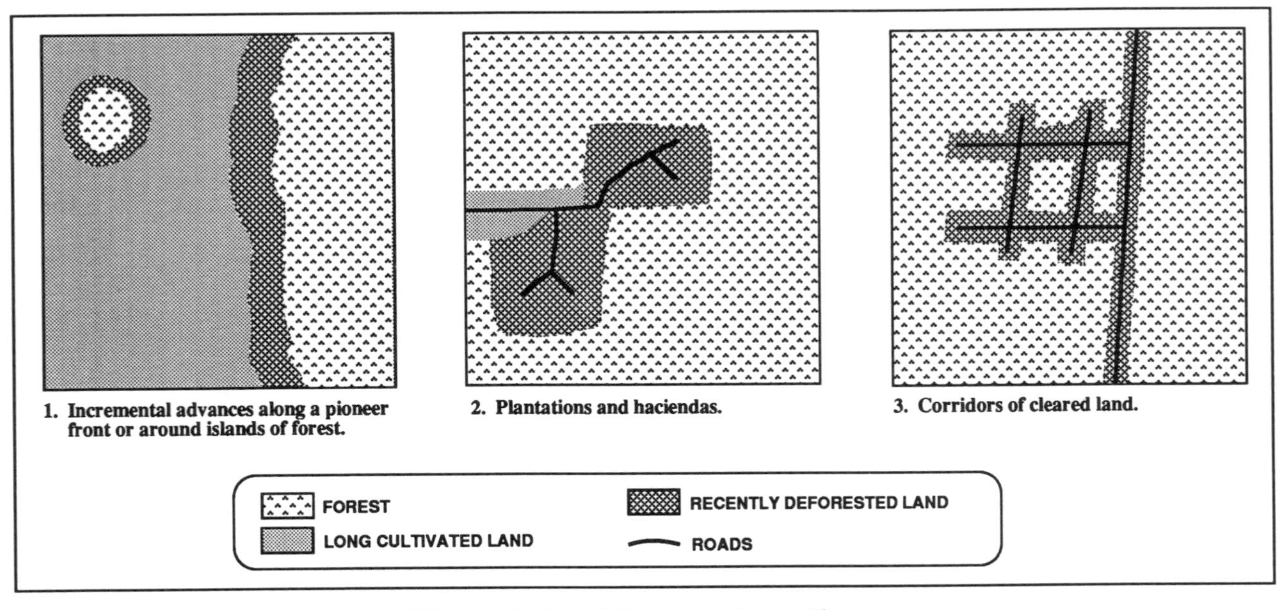

Figure 1.2. Spatial Patterns in Forest Clearing

zonas province farther to the north had only a minimal impact on land use (Smith 1981; Maher 1989:29). As the last two examples make clear, roads only generate rapid deforestation if they provide access to markets. Even after the completion of the roads, Guiviare and Amazonas remained far from major markets, and little economic or population growth occurred. All three examples suggest the contingent nature of tropical deforestation. It only occurs when a certain set of conditions characterize an area.

THE PATTERNS OF CLEARING IN COMPARATIVE HISTORICAL PERSPECTIVE

Differences in the agricultural economies that accompany fringe and corridor clearing suggest that their incidence may have changed during this century. In the few detailed accounts of fringe clearing around large forests, smallholders consumed much of their farm's agricultural production (Keyes 1976). As long as peasants engage in subsistence production, the decline in access to urban markets which occurs when they clear new fields in more remote locations should not constrain them in their search for land. If peasants attempt to market a portion of their crop, they might be reluctant to clear additional land farther into a forest because the costs of transporting the crop to market from a more remote location would reduce their profits. The limited nature of market integration implicit in fringe clearing around large forests suggests that it may have been more common in the nineteenth and early twentieth centuries when more people engaged in subsistence agriculture.

In recent years corridor patterns of clearing have probably become more common in regions with large forests. While a semi-subsistence orientation among smallholders may have facilitated fringe clearing around large forests, a market orientation characterizes all those landowners who clear land in transportation corridors. The linear pattern of settlement in a corridor maximizes the number of settlers who have easy access to roads and markets (Hiroaka and Yamamoto 1980:429). Differences between the two patterns of clearing in the speed with which colonists occupy and clear land underscores the differences in context. While colonists along a forest fringe advance slowly along a long front of activity, colonists who settle along corridors occupy and clear land quickly. They know that the new road has created a market for land and fueled speculative fevers, so they expect that other parties will file competing claims to their land. To discourage their competitors, the colonists work feverishly to clear the land. The clearings secure their claim to the land.

While corridor clearing may have recently replaced fringe clearing as the predominant pattern of clearing in large forests, both types of deforestation are common throughout the world. They usually occur in sequence, with corridor clearing giving way to fringe clearing as deforestation proceeds in a place. Highly capitalized organizations like the state or a timber company begin the process by building a penetration road, and colonists quickly clear a corridor of land along the road. The subsequent construction of feeder roads induces further deforestation and swaths of cleared land appear in the zone, reducing the forests to island remnants away from the roads. Under these circumstances further deforestation occurs in small increments as the poorer inhabitants clear small plots of land in the remaining islands of forest. This type of fringe clearing does not entail a subsistence orientation. These farmers produce for sale in a market.

The changing temporal and spatial coordinates of tropical deforestation give us some preliminary insights into the activities that currently encourage land clearing. The growing importance of corridor patterns suggests that linked decisions by institutions to build roads and colonists to clear land have become increasingly important in the dynamics of deforestation in places with large forests. The widespread and persistent nibbling at islands of rain forest in more densely settled regions suggests that the subsistence strategies of growing families play an important role in small-scale clearing. In this way spatial patterns point to a set of factors that, together, could explain recent changes in worldwide rates of tropical deforestation.

2
A Theory of Tropical Deforestation

■■

THE CONVENTIONAL EXPLANATIONS

Because smallholders decide to clear land before they actually do it, a search for the causes of tropical deforestation should begin with questions about decision-making. What sorts of political and economic dynamics explain the decision of an agency's director to build a road into a rain forest? What kinds of social and economic considerations affect a peasant's decision to start a farm in the forest? A venerable line of analysis with a new name, "political ecology,"offers answers to some of these questions. In this type of analysis differences in political economic interests explain why people choose to clear different amounts of forest.

The political ecology framework appears attractive in part because other theoretical perspectives have so much difficulty explaining decisions to clear land in large blocks of forest. The problems begin with conventional economic theory. It can explain deforestation rates on large landholdings, but it has difficulty explaining deforestation on small landholdings. Increases in prices for agricultural commodities lead to decisions to clear more land in places with large landholdings. When prices rise, landowners build roads into unexploited areas and either log it or clear it for agricultural purposes. In places with small forests, smallholders clear their last patch of forest and plant crops when the prices of agricultural commodities rise. In places with large forests, smallholders may not be able to clear more land because rising prices do not induce the road building necessary to market goods from newly exploited areas. Unlike plantation owners, smallholders do not have the capital to build feeder roads, and the public agencies that build roads in these regions do not benefit directly from an increase in commodity prices. Under these circumstances commodity prices might rise, but additional roads may not get built, and rates of deforestation may not increase.

Other theories focus on the people who clear land rather than the markets in which they sell products. These theories acknowledge that poor peasants play an important role in the process, but they differ in their understanding of the forces that compel peasants to destroy the forests; one explanation emphasizes demographic factors while another explanation focuses on political economic conditions. The demographic explanation has a Malthusian emphasis. In this model growing populations of small farmers create land scarcities that in turn lead to the expansion of agriculture into forested regions (Hamilton 1984; Myers 1984; Stonich 1989:271; Whitmore 1984; Watters 1971:9–12). To accommodate the needs of growing families, some peasants move to distant forests to establish small farms while other peasants cut new fields out of the forest at greater distances from their homes (Keyes 1976; Uhlig 1984). For evidence, analysts point to the widespread emigration of Indonesian peasants from densely populated Java to sparsely populated Kalimantan and Sumatra (Pelzer 1945:232; Scholz 1986:31).

A second explanation for the variable role of smallholders in tropical deforestation asserts that "poverty causes deforestation." Virtually everyone agrees that slow economic growth makes it difficult, if not impossible, for urban economies to absorb all of the outmigrants from rural areas, so some migrants will look for economic opportunities in rain-forest regions. One version of this argument claims that political and economic inequalities in the larger society compel poor, dispossessed peasants to seek a livelihood on the margins of society, in the rain forests of remote, rural regions (Blaikie and Brookfield 1987; Collins 1986; Foweraker 1981; Guppy 1984; Ledec 1985; Painter and Partridge 1989; Plumwood and Routley 1982; Redclift 1989; Schmink and Wood 1987). Sharp, historical inequities in the distribution of land cause large numbers of rural poor to grow up without access to land, and they gravitate to the unclaimed lands on the frontier. Processes of proletarianization swell the ranks of the landless poor and increase the flow of migrants to communities on the fringes of the rain forest (Foweraker 1981:142–150). According to one report,

> People displaced by development projects are often the direct agents of deforestation. While peasant cultivators and herders have done the actual tree cutting and burning, the causes lie in a chain of events that have left these people few options but to destroy the forest or starve. (Office of Technology Assessment 1984:87)

A number of observers have argued for the primacy of population growth (Hamilton 1984; Myers 1984) or political economic factors (Tucker and Richards 1983; Richards and Tucker 1988) as causal agents. These debates about the relative strength of the population growth and proletarianization arguments obscure the degree to which the two explanations complement one another. High rates of natural increase among the rural poor in developing countries result in large numbers of young, impoverished workers entering the rural labor force fifteen years later. The growing numbers of poor workers in settled rural areas reduce wage rates, and this trend encourages the young to move to a pioneer front. Evictions and the sale of small landholdings in regions experiencing agricultural modernization add to the stream of poor migrants headed for the frontier regions. Both the population growth and proletarianization explanations contend that immiserization among the rural poor provides the stimulus for migration to rain—forest regions and the accompanying environmental destruction.

Arguments that attribute tropical deforestation to the actions of the landless poor see the process as never ending. It is a "slow burning fire" (Dourojeanni 1979) that "advances inexorably" (J. Weil 1989) across the landscape, fed by a continuing stream of poor, rural migrants from already settled regions. In the Malthusian model high birth rates generate a momentum in population growth that requires generations to reverse. Even though fertility rates may fall from one generation to the next, the number of women of child bearing age increases, so populations continue to grow, and people continue to push agricultural frontiers outward. The proletarianization argument also suggests that tropical deforestation proceeds inexorably. According to this argument, a comprehensive land reform would reduce the number of landless poor and ease the pressures on pioneer fronts, but enacting and implementing this type of reform is difficult. In the last twenty-five years only Nicaragua has carried out a reform of this magnitude. If these political rigidities play a crucial role in sustaining rapid deforestation, one can expect little change in deforestation rates without a major restructuring of these societies.

Like conventional economic theory, the several variants of the immiserization argument explain important aspects of the deforestation process. Economic theory and Malthusian arguments explain why peasants so frequently clear the few remaining forests in settled regions. As families grow larger and the productivity of the land declines, small farmers without capital convert the last patches of forest on their lands into fields (Collins 1986:3–4). The proletarianization argument explains why frontier towns so frequently become labor reserves in regions undergoing rapid deforestation (Lisansky 1989; Becker 1986; Browder and Godfrey 1990). Landless peasants migrate to development poles in forested areas in search of wage labor and free land. With all of the accessible land taken, the migrants become permanent wage laborers who reside in towns between jobs.

Two empirical problems raise questions about the Malthusian and proletarianization explanations. They err in estimating the pace of deforestation, and they misconstrue who moves to the frontier. As noted above, the immiserization argument portrays deforestation as a steady, inexorable process. In contrast, a number of studies indicate that deforestation proceeds in fits and starts. Remote sensing estimates of deforestation in Brazil during the past fifteen years show sharp increases and decreases from year to year (Woodwell et al. 1987; Malingreau and Tucker 1988).[1] A historical study of spontaneous colonization in the eastern Amazon describes a period of extensive land clearing during the 1970s followed by little clearing during the 1980s (Lisansky 1990:140–165). Rates of deforestation vary considerably from place to place. Bowman (1931:48) points out that pioneer fronts expand rapidly in some places and remain stationary in other places. Under some circumstances pioneer fronts will retreat. During the 1950s landslides forced the closing of the penetration road into Satipo in the Peruvian Amazon for ten years. The closing of the road led to farm abandonment and an exodus from the region (Shoemaker 1981:81—92). Other studies of land use in diverse places such as northeastern Peru and central Africa show little change in forest cover (Rhoades and Bidegaray 1986; F.A.O. 1981a).

An extensive review of the literature on migrants to rain-forest regions suggests that both the Malthusian and proletarianization explanations underestimate the resources that most migrants bring to the frontier. The proletarianization explanation, with its emphasis on the contributions of the *colono* system to deforestation, provides the most complete account of how impoverished individuals might deforest large areas, but this explanation overestimates the willingness of peasants to start farms in rain forests. This miscalculation and the problems with Malthusian theory may stem from a common source, a failure to appreciate the difficulty of opening up large blocks of rain forest for settlement.

Malthusian and proletarianization theorists generally ignore the variable natural features of rain-forest environments; they seem to assume that rain forests represent a uniform landscape that a poor person with an axe or a chain saw can bore into indefinitely. Uneven terrain, poor drainage, or impassable rivers rarely receive any mention in these reports. This benign image of the rain forest may stem in part from the recent flood of writing by ecologists about rain forests. As Wallace noted long ago (1853:309), naturalists underestimate the hardships that rain forests impose on poor residents or travelers. As the quote in the frontispiece suggests, psychological burdens amplify the physical challenges of living in the shadows of the forest. A failure to appreciate these physical and psychological difficulties explains why social scientists so often err in identifying who clears the land and when they do it. The following sections explore this link between the pace of deforestation, the characteristics of the people who do it, and the physical setting in which they work.

THE PACE OF DEFORESTATION: THE COLONO SYSTEM IN LATIN AMERICA

The literature in rain-forest destruction contains numerous references to an arrangement between landlords and peasants, known as the *colono* system in Spanish speaking countries. In this arrangement poor peasants do the initial trailblazing and clearing of the land; after several years of cultivation, crop yields begin to decline, and the peasants turn the land over to a wealthy class of landlords. The peasants then move farther into the forest, stake out a claim to some unoccupied land, and begin the cycle over again. The wealthy landowners convert their recently acquired lands into cattle pastures, creating a "hollow" frontier, a place that has become depopulated now that large cattle ranches have replaced the subsistence plots of the pioneers (James 1959:491–492). Because the agrarian class structures that sustain this pattern of settlement and dispossession characterize entire frontier regions, the *colono* system could explain the advance of lengthy pioneer fronts into the forest. This pattern of change in tenure and land use reportedly exists in a wide variety of Latin American places, in Brazil (Wood 1983:266–267; Foweraker 1981), Bolivia (Painter and Partridge 1989; Stearman 1985), Colombia (Ortiz 1984), Venezuela (Watters 1971), Central America (Parsons 1976; Jones 1989), and southern Mexico (Price and Hall 1983). In places where settlement occurs in corridors along roads or rivers the *colono* system could explain the widening of the corridors. They grow wider as settlers lay claim to and clear land at greater distances from the road or river. The logic of the *colono* system suggests that a corridor of cleared land could continue to widen indefinitely. New arrivals set themselves up in the last row of farms, five or ten kilometers from the road and begin to clear land. The wealthier farmers close to the road exploit the new colonists. Eventually the colonists sell their land to the more established farmers and move farther into the jungle to start the cycle of deforestation and economic exploitation over again. With a steady stream of poor farmers entering colonization zones in search of land and other colonists trying to establish farms for a second or third time, the process of economic exploitation and deforestation could continue indefinitely.

Although commentators frequently refer to the *colono* system, few people have described it in detail. The most detailed account of the *colono* economy comes from a study of colonization along rivers in Caqueta, Colombia (Roberts 1975). The Caqueta study documents the economic exploitation suffered by the colonists, but it raises questions about the amount of deforestation that accompanies the process and the degree to which the process reproduces itself by driving failed colonists deeper into the forest (Roberts 1975). Day to day problems of subsistence afflict the colonist in the last *respaldo* (row). For several years after establishing a claim, most of his land remains in forest, and the farm produces very little revenue, so he must support himself and his family by working for wages on the farms of more established colonists. The off-farm work reduces the amount of work that the colonist does on his own farm, limits the amount of land he clears, and decreases the income he earns from the farm. The more established colonists with farms closer to the road have larger incomes from their farms, so they do not hire themselves out to other farmers; instead they set up reciprocal labor exchanges with one another. In return for helping a neighbor with a particular task, the established colonist receives his neighbor's help with a task on his farm.

The colonist in the last *respaldo* encounters other obstacles when he begins to market his crops. Because he has little capital, he usually does not own either a mule or a horse, so he has to rent pack animals from a more established colonist in order to transport his crop to market. If a truck owner and a new colonist do not have an arranged time to meet, which is often the case when the colonist only has small amounts of produce to sell, the colonist may have to sell his crop to another established colonist who lives along the road or river. The colonist/middleman will in turn sell the crop to the owner of a passing launch or truck. The expense of paying for transport to the road, coupled with the low prices paid for products by local middlemen, leaves the colonists in the last row with little profit.

During the colonists' first few years in the forest they pile up debts in an effort to develop their farms. Nearby store owners, many of whom are more established colonists, allow the new colonists to purchase foodstuffs on credit with the understanding that the newcomers will repay their debt in labor. This requirement slows down the development of the new colonists' landholdings. Banks will sometimes provide credit to the new farmer, but only if an established colonist, with valuable collateral, cosigns the loan (Roberts 1975:195). This agreement between established and new colonists usually costs the latter dearly.

Local variations in soil fertility can aggravate or ameliorate the colonist's economic predicament. Where the soils are fertile, good harvests may prolong the colonists' stay on the land. More frequently, the soils lose their fertility rapidly, causing what Bolivian colonists call a *barbecho* crisis (Eden 1990:107–110, 113–118; Maxwell 1980; Stearman 1985:158). A succession of poor harvests, coupled with the slow pace of land clearing, high marketing costs, and mounting debts, make it impossible for new colonists to progress. To escape their predicament, colonists sell their partially deforested farms and move elsewhere.

Where do they go? According to proletarianization theorists, dispossessed peasants move deeper into the forest and begin clearing land all over again. What Bowman calls the "invitation of the land" (1931:38) makes decisions by poor peasants to embark on a new cycle of farm development and forest destruction seem somewhat credible. The unoccupied lands may not be too distant, ten or twenty kilometers away, and the procedures for establishing a claim, clearing the land, and sowing the crops are familiar, so the challenge of carving a new farm out of the forest may not seem so daunting. Other considerations make a decision to start over seem unlikely. First, a move farther into the forest would reduce the number of crops that the colonist could grow profitably (Dunn 1954; von Thunen, 1851). For colonists who had just lost a farm because they could not market their crops, the prospect of failure in another remote location must seem very real. Second, the punishing physical task of clearing primary forest would discourage many older colonists; "few people try it twice, very few men over forty try it, and fewer still try it three times" (Redclift 1987:126). Third, a family that has been forced off its land or sold it for a low price after some disappointing harvests will not have the savings to invest in a new landholding. How will the family support itself for the eight to ten months until the first harvest?

So where do recently dispossessed peasants go? In his Colombian study Roberts indicates three destinations. Some people moved to the provincial capital to work as wage laborers; others returned to the Andean highlands to work on haciendas, and a large number of the dispossessed peasants moved farther down river, presumably to start over again. Because this last group could be construed as moving deeper into the jungle in a manner consistent with the proletarianization argument, their move requires special scrutiny. By moving down river, colonists could presumably acquire land on the river. Although they might be farther from urban centers, their new location would eliminate the cost of transporting crops to the river, and this decline in costs would more than offset the increased costs of transporting goods upstream to urban markets in the Andes. Where roads rather than rivers are the means of transportation, a colonist would start over in a new location farther from the road if there was some hope that over time the construction of a road would bring him closer to markets than he had been in his previous location. The *colono* would pioneer in a new location with the expectation that this time he will succeed where he previously failed. When large numbers of failed colonists choose to start over again in new, more remote locations, deforestation does advance in an inexorable fashion but, as the examples make clear, failed colonists usually do not choose to colonize anew unless someone else has decided to build roads into a region. In the absence of new infrastructure such as the construction of feeder roads out from a river or a main road, colonists will not push deeper into the jungle in order to acquire land for farming. In other words rain-forest corridors do not widen inexorably, even in places where large numbers of peasants have no land.

The *colono* system would produce continued deforestation if, as reported in a study of colonization in Costa Rica (Sewastynowisc 1986), some colonists sold their lands at a profit. Then they might have the resources to pay someone to help them clear land on their new farm. Because land values increase with the improvements in access provided by newly constructed roads, it is not unusual for the value of smallholders' land to appreciate. Presumably peasants who choose to trailblaze a second time do so in areas toward which a road is being constructed. Having capitalized on the increased value of their first farm as a road approached, peasants may want to repeat the cycle. This situation seems far removed from the *colono* system with its emphasis on exploitative relationships between the classes.

The *colono* system might work as described if dispossessed peasants have no other choice but to begin carving another farm out of the forest. Hiroaka, in describing how Bolivian peasants cope with the *barbecho* crisis, suggests that they do have choices. They can work for wages in their present locale, depart for the cities to work as wage laborers, or begin again on another colonization front (1982:96). Evidence from rain-forest frontiers in Peru (Aramburu 1984), Venezuela (Watters 1971), Brazil (Becker 1986), Colombia (Roberts 1975), and Nepal (Shresta 1989) suggests that this array of choices is not unusual. One might hypothesize that the poorer of the dispossessed peasant households would opt for wage labor either locally or in cities while the better off peasants might move farther into the forest. This line of reasoning suggests that "success breeds success" in tropical deforestation; peasants whose initial investment in the development of tropical lands yields high returns may try it again, particularly if appropriate conditions, meaning the construction of a road toward a sector of jungle they know, present themselves. Poorer peasants would opt for the more certain return of wage labor, especially in the absence of plans for the construction of roads farther into the jungle. If these suppositions prove true, the resulting pattern of colonization and forest destruction suggests that peasants only create a series of farms over the course of a lifetime when they can take advantage of road building and other regional development projects initiated by large institutions. By implication deforestation proceeds irregularly, following the ups and downs of road building in a region.

POVERTY AND DEFORESTATION: MIGRATION TO THE FRONTIER

Immiserization theorists do not know who clears most of the land. As Moran (1987:80) has pointed out, social scientists have usually characterized colonists as a "lumpenproletariat" who have migrated in response to victimization by a capitalist class of large landowners. There are empirical grounds for questioning this characterization. The poorest of the poor usually do not migrate to frontiers (Raison 1981:62). They are too absorbed in the compelling business of daily survival to investigate fully the opportunities offered by settlement in a distant place. In this view colonization represents an investment by individuals, and its absence where natural conditions permit it suggests a condition of underdevelopment. Prior to the recent surge of deforestation and development, the world's tropical zone had both starving populations and virgin lands. Northeastern Brazil had "people without land" living near a "land without people" in the Amazon basin.[2] In the words of P. Monbeig, a French geographer, "the weakness of pioneer advance where it should be vigorous . . . is a consequence and a form of underdevelopment" (1966:1004).

Although much of the literature on migrants to tropical frontiers focuses on participants in state sponsored colonization schemes, 75 to 80 percent of the people opening up new lands for settlement around the world are spontaneous rather than state sponsored colonists (Scudder 1981). Because arid zones with their expensive irrigation schemes contain disproportionate numbers of state sponsored colonists, the proportion of spontaneous colonists in the humid tropics may be even higher (Scudder 1981:245). They are the chief agents of forest destruction in rain-forest regions. Spontaneous colonists are not "the poorest of the poor" but come predominantly from peasant families that have some resources.[3] Because well-to-do peasants move in response to the pull of economic opportunities rather than the push of economic hardships (Connell et al. 1976:19), surveys of colonists in rain-forest regions should reveal an investment orientation. Migrants who carved cocoa farms out of forests in central Ghana between 1890 and 1930 exhibited this orientation.

> The essential nature of the migratory process is . . . forward looking, prospective, provident, prudential—the opposite of hand to mouth. They [the migrants] did not eat the proceeds from their early farms. Rather, almost from the beginning the farmers regarded themselves as involved in an expansionary process from which they had no intention of withdrawing. . . . cocoa farms . . . were regarded as investments—property that existed for the purpose of giving rise to further property. (Hill 1963:179–180).

Surveys of Philippine migrants to the forested valleys of southern Mindinao indicate an investment orientation (Simkins and Wernstedt 1971:80). Even spontaneous colonists in Indonesia, faced with extreme land scarcity in their place of origin (Java), emphasize pull rather than push factors in explaining their decision to migrate. As one expert put it,

> Unassisted transmigration from Java is essentially a response to opportunities to obtain a living in another land . . . there is no correlation between poverty and the decision to migrate. This is even more the case with spontaneous than with government supported transmigrants. It is the wish to improve one's economic position rather than overwhelming poverty that encourages unassisted transmigration from Java. (Hardjono 1985:53).

Pioneers take large risks when they claim land by blazing trails in a rain forest (Eckstrom 1979:36; Scudder 1981:94; Sewastynowisc 1986:749; Moran 1989:26; Pelzer 1945:110, 245). The colonists must find a way to provide for their families while they invest six or eight months in clearing land and planting crops. Even after six months' of labor, the colonists' investments are not secure because they do not have title to the land. Diseases like malaria add to the risks colonists face. Under these conditions, only colonists with financial reserves will venture unassisted into a frontier region. Because the immiserization model does not fully acknowledge the difficult circumstances facing colonists in rain-forest regions, it does not seem plausible. People who have just lost their land or who have grown up without land do not have the resources to undertake trailblazing. Poor people do move to pioneer fronts, but they only do it under the auspices of a more powerful person or institution. Immiserization theory does not appreciate the importance of these combinations of capital and labor in land clearing.

THE SOCIAL ORGANIZATION OF TROPICAL DEFORESTATION: GROWTH COALITIONS, LEAD INSTITUTIONS, AND FREE RIDERS

The validity of the above critique depends on the size of a rain forest. When a forest is small and located near an established settlement, population growth or proletarianization can explain most deforestation. Driven by subsistence needs, landless sons or dispossessed peasants clear small tracts of forest near their homes. In this setting land clearing does not require the extensive resources that it often requires in largely forested regions. In most instances large blocks of rain-forest are inaccessible; mountain ranges, impassable rivers, or remote locations make it costly to settle in these places. The landless peasants featured in the immiserization argument recognize the obstacles and do not attempt to settle in these areas unless they receive extensive assistance. In other words large rain forests only undergo extensive settlement and deforestation when large numbers of people tied together in some form of complex organization work together to open up a forested region for development. The two forms of organization that facilitate the destruction of large forests are outlined below.

Growth Coalitions

In many instances coalitions of actors from different social classes open up regions for settlement and deforestation. The organizers vary considerably. They can be wealthy relatives, urban investors, or government colonization agencies. Their support diminishes the risks of pioneering to the point where a poor colonist will undertake it. For example, colonization agencies often provide settlers with housing and food for the first growing season (Scudder 1981). Young workers without resources move to frontiers when they know that they can secure work, food, and shelter upon arrival. In one frequent arrangement, a peasant will work for a large landowner in a pioneer region until the peasant has accumulated some savings and identified a tract of unclaimed land in the area. At that point he begins to work part-time for the landowner and devotes the remainder of his time to clearing land on his new farm (Collins 1988:182–184; Findley 1988:277). Sometimes the colonist's initial arrangement involves working for relatives who settled in the zone at an earlier date (James 1983:581). The affluent landowner benefits from these arrangements because they provide him with labor to develop his landholding.

Poverty is not the only basis for seeking alliances with others. Relatives and neighboring landowners with similar amounts of money also form "growth coalitions" (Molotch 1976; Logan and Molotch 1987:32–34). By forming coalitions, smallholders are able to amass enough resources to overcome obstacles to settlement. With their pooled resources they can open up lands on the far side of a fast-running river by purchasing a cable to string over the river, or they can secure their claim to a tract of land by paying someone to stay on the land at all times. If the colonists encounter resistance from indigenous peoples, their ability to wrest control of the lands from native peoples depends on their ability to mobilize the coalition's resources.

Coalition members usually divide up the work according to the resources they bring to the coalition. Government officials lobby for public funds to build a road into the region, landless peasants take on the arduous job of trailblazing, and wealthy merchants provide funds which sustain the peasants in the forest. While everyone in a coalition hopes for speculative gains in the value of land, the number of members with purely speculative interests varies from coalition to coalition. Groups with many speculators sometimes secure infrastructure quickly, but they clear land slowly. Groups with fewer speculators demonstrate different proficiencies, clearing land quickly but constructing roads slowly, if at all. In all of these coalitions the members "profit through the increasing intensification of land use in the area in which they have a common interest" (Molotch 1976:310—311).[4]

The coalitions take different forms. Colonization programs bring peasants and government officials together in a formal alliance to clear the forest. In some instances peasants form kin or locality based coalitions to facilitate the development of a tract of land. In other instances colonists working neighboring tracts of land form loose coalitions to advance a common project such as the construction of a feeder road. The following description of the different types of coalitions focuses on the working relationships that tie coalition members to one another.

Colonization Programs. Private and state sponsored colonization schemes bring colonists and lead institutions together in formal ways. The institutions provide roads or rails for transportation, subdivide the land, and in some instances provide housing. Each country in the Amazon basin and four countries in Southeast Asia have established colonization programs to convert large areas of forest to agricultural uses (Scudder 1981).[5] In Latin America, particularly in southern Brazil and Paraguay, private colonization companies have developed large tracts of forested land for agricultural use (Margolis 1973;

Muller 1974). By providing basic services in pioneer settings, public and private organizations reduce the risk associated with pioneering and make settlement in government assisted communities an attractive option for poor, dispossessed peasants. These settlements in turn become magnets for large numbers of spontaneous colonists (Bilsborrow 1991; Martine 1980). Government officials do not create colonization programs out of the goodness of their hearts. In democracies they receive tangible political benefits from the programs. Having received a wide array of special government services, the residents of new communities usually vote for political incumbents. The size of this vote depends on the number of new communities that the regime has established in recent years, so local and federal officials have a powerful political incentive to establish as many new communities as possible (Fearnside 1989:214–215). Colonization programs also give the government favorable press releases in urban centers because, by putting land into the hands of poor peasants, they reduce the pressure for politically disruptive land reforms (Domike 1970).[6]

Coalitions of Kin or Friends. Relative political and socioeconomic equality characterizes the members of kin and locality based coalitions. Families in Gima, a small community on the eastern edge of the Ecuadorian Andes, cooperated extensively between 1940 and 1975 in settling the Rio Cuyes region, an adjoining river valley on the eastern slope of the Andes. Relatives from Gima settled on adjacent tracts of land in the frontier region and cooperated extensively in the construction of a network of trails. During the 1960s and early 1970s the Rio Cuyes colonists joined together to lobby the regional development agency for assistance in building a road from Gima into the frontier zone. In addition the colonists periodically pooled their labor to accomplish important tasks in the construction of the road (Eckstrom 1979). Hmong kinsmen in northern Thailand usually migrate in groups of two or more families when they decide to open up a distant area for cultivation. These families provide one another with mutual assistance of all kinds (Geddes 1976). Between 1890 and 1930 hill farmers in central Ghana formed "companies" in their highland communities in order to purchase and develop cocoa farms in the adjacent lowlands. As soon as the companies acquired land, they disintegrated; the farmers reactivated their coalitions twenty years later when they decided to build bridges over several lowland rivers in order to market their crops via a newly constructed railroad (Hill 1963:30–74, 232).

Some coalitions form after the migrants have arrived in the humid lowlands. Co-workers from Bolivian plantations have repeatedly formed syndicates and occupied tracts of rain forest, which they then subdivided among themselves (Gill 1987:76). Neighboring colonists in Colombia's coastal lowlands often come from different highland provinces, but they have joined forces to lobby for the construction of roads in the lowlands (LeGrand 1986:26).

Patron-Client Ties. Probably the most common type of growth coalition involves an affluent individual or institution that provides enough resources to establish a "bridgehead" settlement in an area (Findley 1988:277). Poor colonists follow a leader or lead institution into the forested region. The relations between these individuals vary widely in their content, with many of them approximating traditional patron-client relations. The patron gets a daily good, wage labor at below market prices; in return the client gets an extraordinary, difficult to evaluate good, knowledge about a region with numerous economic opportunities. In another arrangement, common in the large areas of Southeast Asia where peasants practice commercial shifting cultivation, trader/creditors help indebted peasants acquire land for a reduced price; in return the peasants agree to sell their harvests and after several years the land itself to the trader. He then resells the deforested land to another family of settlers. Some of the original settlers earn enough from the first sale of the property to buy their second farm without the assistance of traders (Uhlig 1984:161–163).

Unequal exchange usually characterizes the ties between patrons and clients. Scholz (1986) describes one arrangement in eastern Thailand in which peasants clear a tract of forest every two years and plant sugarcane for the owner. For the two-year period the peasants live off of crops, which they interplant with sugarcane, and then they begin again with a new clearing. Should foresters try to enforce the law against the destruction of forests, the peasants, but not the landlord, are liable for the destruction of the forest. The landlords will, however, attempt to block prosecution by paying a bribe. Substantial numbers of peasants in Southeast Asia escape from these arrangements after several years and succeed in establishing themselves as independent producers on the frontier. This type of coalition occurs in many places, including Panama (Jolly 1989), Peru (Collins 1988), Bolivia (Gill 1987), Thailand (Scholz 1986), and the Philippines (James 1983). Unlike the *colono* system, patron-client ties reduce the risks of trailblazing and, in so doing, make a move to a frontier feasible for impoverished peasants.

Patron-client relationships between missionaries and colonists spurred the settlement of several rain-forest regions in Latin America (Ortiz 1984:207; Marsh 1983:29; Sarmiento Chia 1985:67; Weil and Weil 1983:32–33). The creation of missions with schools and medical dispensaries reduced the risks facing potential migrants. The colonists supported the missions, and the missionaries spent their resources promoting the construction of roads into the colonists' communities (Ortiz 1984). Some of these coalitions endure for decades while others, especially the loose coalitions between neighboring colonists, last only a few months.

The alliances that open up regions for deforestation are often more tacit than explicit. In many instances the lead institution and the surrounding peasant population develop a symbiotic relationship, but never act together in pursuit of a common good. The transnational fruit companies, which made the first extensive clearings in the rainy Caribbean lowlands of Costa Rica and Colombia, attracted landless peasants to the new plantations by offering employment, building roads, and constructing public health facilities. Once in the region, some of the companies' workers left the plantations, established their own small farms in close proximity to the plantations, and grew foodstuffs for the plantations' workers (Parsons 1976:122; LeGrand 1984:192–193). A similar sequence of events occurred during the 1930s around newly created plantations in the southern Philippines (Simkins and Wernstedt 1971:43–48).

Lead Institutions and Free Riders

Large organizations, like oil companies, often take unilateral action to open up a region for development, and small farmers follow the companies into the region. To exploit newly discovered mineral deposits or extract well known timber resources, companies build penetration roads into forested regions. The companies do not try to control access to the roads, so colonists claim the roadside land, clear it, and use the roads to get to market. In this respect the colonists are free riders; they make heavy use of a collective good (the road), which they did nothing to provide.[7] In one sequence of events, documented on all three continents, loggers construct a network of logging roads and extract the most valuable timber from an area. Peasant cultivators take advantage of the improved access offered by the logging roads and establish small farms along the roads (F.A.O.-U.N.E.P. 1981a; Kartawinata and Vayda 1984; Vayda and Sahur 1985; Brownrigg 1981:317–318; Fearnside 1991:199). In preparing the land for cultivation, the small farmers complete the job of forest clearing begun by the loggers. In a similar sequence of events mineral companies, eager to develop new deposits, or the state, anxious about sovereignty over a border area, build penetration roads into a rain-forest region, and free riding peasants settle in a corridor along the road.

The opening up of a region through the construction of a penetration road often has dramatic environmental consequences because no one has firmly established property rights to the forested lands along the road. Governments claim large blocks of rain forest as property of the state, but they allow their citizens to settle on the land, so in effect these lands are free goods (Hazlewood 1986:78–82). When lead institutions construct penetration roads that provide easy access to these lands, individuals and interest groups scramble to appropriate the free good; they work feverishly to establish their claim by clearing a portion of their land. While the initial clearings may constitute only 10 percent of the claimed area, they account for much of the initial surge of deforestation that follows the construction of an access road. The presence of more unclaimed land just over the horizon at the head of an advancing road discourages attempts to conserve resources, so the clearing of land and the mining of the soil continues unabated (Hazlewood 1986; Redclift 1987:142; Sedjo and Clawson 1983; Margolis 1973:220–230).

Under these conditions the settlement process takes on aspects of a game. Peasants, local elites, and urban investors try to capitalize on the construction of the next penetration road by predicting the road's route and laying claim to land in its path (Watters 1971:9–12; Scholz 1986:17–19). While peasants and speculators frequently form groups to occupy lands and contest competing claims (Uhlig 1984:240–241), smallholders do not have to pool their resources to the same degree as they would elsewhere because, with the transportation link in place, they do not have to mount an effort to improve access to the region. In sum when a lead institution such as a mineral company builds the infrastructure, growth coalitions play a smaller role in the subsequent claiming and clearing of land.

The mix of growth coalitions and lead institution—free rider combinations that prevails in a place also depends on local variations in the organization of work and the incidence of class conflict. Where plantation agriculture prevails, labor requirements bring planters and peasants into close association and the contact facilitates the formation of cross-class coalitions to open up new regions for settlement. Where timber companies open up lands, the loggers usually do not hire large numbers of local laborers, so they do not come into regular contact with the farmers who cultivate lands along the logging roads. The segregation of the two groups discourages the formation of growth coalitions and encourages aspiring colonists to look for free rider arrangements.

The incidence of class conflict also affects the way people organize to clear land. In places like southeastern Ecuador the relatively low levels of class conflict do not inhibit the formation of cross-class alliances, so growth coalitions flourish. In places rife with class conflict like the parrot's beak region of eastern Brazil, high levels of class conflict may discourage the formation of growth coalitions. In this setting, lead institution-free rider arrangements account for more land clearing than they do elsewhere. Government policies contribute to these differences in the incidence of class conflict. The Brazilian government has traditionally made large grants of land to absentee owners (Branford and Glock 1985), which in turn has touched off conflicts between "owners" and peasant squatters. The Ecuadorian government makes large grants, but they only go to consortia who have already established a presence on the land. This practice minimizes violent conflict because large landowners confront peasants early in the settlement process when they can be bought out more easily.

Regional development policies also contribute to the different levels of class conflict. The *Polamazonia* program, initiated in 1974 in the Brazilian Amazon, funnels tremendous amounts of capital into fifteen designated growth pole regions (Hall 1989). By pouring government funds into the development of small areas, the state encourages aspiring industrialists and small farmers to compete for lands near the growth poles. In this way the state sets the stage for intense competition and conflict between the two parties. In the aftermath of disputes former antagonists do not cooperate, but they move quickly to exploit newly discovered resources or take advantage of new infrastructure. In this setting, more so than in other places, surges in deforestation await the initiatives of leaders, either large landowners, corporations, or the government. In Ecuador the continuing influence of regional elites insures that government resources get spread fairly evenly across the provinces, so growth poles have not become as important a focus for migration and investment as in Brazil, and the competition for land, perhaps because it is less intense, has not touched off as many violent disputes. Accordingly, growth coalitions should form more frequently in the Ecuadorian Amazon than they do in the Brazilian Amazon.

These differences in patterns of activity do not remain frozen in time. The opening of a road prompts land speculation and land grabbing, which increases conflict in a place. Land values stabilize as the advance of the road takes the construction crews into other areas (Eckstrom 1979:146–147). The owners then turn their attention to earning an income from the land, and levels of conflict decline. In the calmer atmosphere, cross-class coalitions to claim tracts of rain forest form more easily.

CAPITALIST EXPANSION, LEAD INSTITUTIONS, AND TROPICAL DEFORESTATION

Peasants intent on claiming land in a rain forest face a series of strategic choices. They have to assess the quality of the land in a colonization zone. Will the land beneath the forest support intensive agriculture for a number of years? To answer this question, young peasants usually tour a colonization zone before they decide to claim land there. Expectations about the actions of other interested parties also figure in the decisions that peasants make (Long 1958:253). Will the more powerful members of the coalition take all of the most productive land for themselves? Will the other coalition members have enough resources to make good on their promise of improving access to the region through the construction of a road or an airport? Will the proposed route of the new road take it through the lands a peasant has claimed? When will the road arrive? If a peasant makes optimal choices, the construction of the road provides him with a windfall profit in the form of easy access to markets and an increase in the value of his land. Of course, if arable and potentially accessible lands do not present themselves during a tour, peasants may choose not to colonize new lands. Because the extent of recent road construction in a region determines in large part the supply of accessible, arable land, one crucial factor in a colonist's thinking has a macroeconomic origin. In this sense macroeconomic factors shape individual decisions which, in the aggregate, determine the rate of deforestation in smallholder regions.

If initiatives by lead institutions play a crucial role in opening up large blocks of primary forest for destruction, then the rate of deforestation depends on the frequency with which lead institutions undertake initiatives. Companies start new projects in response to national and international economic expansion. The developed countries' growing appetite for raw materials has induced multinational corporations to launch ever more persistent searches for deposits of oil and minerals in rain-forest regions. A similar increase in the demand for wood products has contributed to a sharp increase in logging in the forests of Southeast Asia and West Africa. In this way growth in the world economy has accelerated processes of deforestation since World War II.

The growth in international trade has also strengthened the state in developing countries. With limited tax bases and systems of tax collection governments found it difficult to raise revenue before World War II (Migdal 1988:8), and they made only feeble efforts to develop peripheral places. Other organizations made occasional efforts to build roads in these places. Foreign companies received permission to build railroads in northern Bolivia early in the twentieth century (Hennessy 1978:100). In southeastern Colombia a wealthy family built a network of roads during the 1930s and 1940s (Roberts 1975; Marsh 1983). In recent years taxes paid by companies extracting oil or minerals have eased the revenue problems of some governments.

Some of the new funds have financed projects in peripheral, rain-forested regions.

When extractive enterprises in rain-forest regions begin to pay taxes, state managers realize the value of natural resources in these places, and they move to secure their nation's sovereignty over these regions. Governments usually try to secure their sovereignty over peripheral places through development. During the 1970s and 1980s governments offered subsidies and tax holidays to large corporations if they invested in peripheral places like the Amazon basin. Government programs sometimes serve the same purpose. Colonization programs that populate an area with loyal citizens establish national sovereignty over an area by creating "living" frontiers. Programs to eradicate malaria from tropical zones such as the Terai in Nepal remove an important obstacle to establishing settlements in sparsely settled areas (Caldwell 1990; Nicolai and Lasserre 1981). The extension of police power into peripheral areas reduces the use of forested areas as buffer zones between hostile groups and, in so doing, encourages settlement in these areas (Castro 1988). By making investments in peripheral places more secure, these programs encourage the formation of growth coalitions that develop and deforest the land.

Changes in international finance have contributed indirectly to more rapid rates of deforestation. Following World War II, the western economic powers promoted international flows of capital by creating four multilateral landing institutions—the World Bank and three regional development banks: the Interamerican Development Bank, the Asian Development Bank, and the African Development Bank. These banks raise money in capital markets and loan it to third world governments, sometimes at concessionary rates, for dams, roads, bridges, and other infrastructure. Bilateral development assistance from wealthy nations has supplemented the funds from the development banks. The United States Agency for International Development (USAID) has frequently lent funds to countries for infrastructure projects.

International lending agencies often provide the financing for projects in peripheral places. Development projects in rain-forest regions, with their emphasis on the construction of basic infrastructure like penetration roads, fit the banks' definition of an appropriate use for their funds. A variety of lending institutions have made this type of loan in recent years. The World Bank provided financing for the Transmigration Program in Indonesia, the Polonoreste Project in Brazil, and the Caqueta Colonization Program in Colombia. Regional banks like the Interamerican Development Bank have financed large numbers of penetration and feeder roads in the humid tropics (Nelson 1973:86, 131). Finally, bilateral aid programs have also provided funds for road construction. USAID financed the road construction into the Alto Beni colonization zone in Bolivia during the 1960s (Nelson 1973:86). Most recently, the Japanese government offered to build a new penetration road connecting Pacific ocean ports in Peru with the Brazilian province of Acre in order to stimulate development in the eastern portions of the Brazilian Amazon.[8] In some instances these roads stimulated the formation of growth coalitions to develop land; in other instances free riders claimed and cleared land along the future path of a road.

The number of road construction projects undertaken by lead institutions increased between 1950 and 1990 with capitalist expansion and the strengthening of the state, but the increase in projects occurred cyclically, following the ups and downs of business cycles. Loans from banks and investments in infrastructure by large companies increase during prosperous periods and decline during recessions. Increases in crop prices during booms spur the coalitions' efforts at agricultural expansion. Extraordinary expenditures or lines of credit for infrastructure spur road building, which in turn stimulates rapid land clearing. During busts falling agricultural prices and declines in funds for road building lead to a sharp drop in rates of deforestation. Roads fall into disrepair (Allen 1975; C. Weil 1989:265); access to markets deteriorates; growth coalitions collapse, and deforestation slows.

Changes in a country's debt burden also affected rates of deforestation during the 1970s and 1980s. When developing countries took on large amounts of debt during the 1970s, they spent liberally for infrastructure, and rates of deforestation increased. When the growing debt burden caused fiscal crises during the early 1980s, agencies cut back on road building, and rates of deforestation slowed. Inman (1991) finds support for this logic in a cross-national study of deforestation during the 1970s and 1980s. Increases in public debt produced short-term increases in deforestation rates. In contrast, a large public debt during the 1970s predicted lower deforestation rates a decade later as the economic crisis precipitated by the debt reduced a state's ability to finance additional road building.

CHANGING PATTERNS OF DEFORESTATION

The postwar increase in the construction of penetration roads suggests a way to reconcile the immiserization and growth coalition—lead institution arguments about tropical deforestation. Before 1960 neither the state nor the mineral companies constructed many penetration roads in frontier regions, and deforestation occurred primarily through the incremental advances of colonists along pioneer fronts. In regions with highly unequal patterns of landholding, agrarian elites loomed large as a source of income, and they wielded considerable power over the rural poor. The small urban labor markets did not offer an alternative source of employment, so sizable numbers of dispossessed small farmers may have tried again and again, in a pattern consistent with the *colono* system, to carve new farms out of the forest as tenants of more affluent families. Not surprisingly, these small farmers moved many times during the course of a lifetime on the frontier (Roberts 1975; Price 1990). In places where smallholders owned most of the land, population growth from generation to generation may have determined the pace with which frontiers advanced.

After 1960 several important aspects of this situation began to change. The increased construction of penetration roads made corridors of cleared land rather than pioneer fronts the predominant form of deforestation in large blocks of rain forest. The new penetration roads offered rich and poor alike a chance to capitalize on the increased land values near the road. The prospect of these roads spurred the formation of growth coalitions to lobby for a particular route, expedite the construction of the road, and develop the nearby lands. The rapid expansion of urban labor markets during the same period provided landless sons and dispossessed peasants with an alternative to starting over again with another farm deeper in the jungle. Under these circumstances peasants only settled in large forests when they could pool their resources or draw upon the resources of a lead institution in developing their lands. The changed pattern of peasant decision-making reflects the growing availability of capital and the expansion of urban labor markets in the second half of the twentieth century. Under these conditions, the causes of tropical deforestation have changed. In the large blocks of forest in the Amazon basin growth coalitions and lead institutions have replaced the *colono* system as the primary engine of deforestation.

THE THEORY AND THE CASE STUDY

The preceding discussion begins by distinguishing between deforestation in large and small forests. When the large forests become small forests, the array of causal forces should change. Growth coalitions and lead institutions start the destruction of large forests. Population growth, proletarianization, and rising agricultural commodity prices continue the destruction in the now smaller, more fragmented forests. Figures 2.1 and 2.2 outline this argument and the widely accepted immiserization argument. To clarify questions about causal order, each diagram distinguishes between trends in the larger context and participants in the process. The trends create conditions that persuade individuals and groups to clear land. Figure 2.1 presents the immiserization argument. Path #1 in this figure represents the contributions of logging operations, mining companies, and large cattle ranches to deforestation. Path #2 describes how growth in the numbers of impoverished rural peasants, through population increase and proletarianization, results in widespread land clearing. Figure 2.2 presents the growth coalition—lead institution argument. Path #1 represents land clearing by lead institutions and free riding smallholders. Path #2 denotes the contributions of the coalitions to deforestation.

The two models through into relief an important difference in the way observers see the contributions of economic classes to deforestation. The immiserization model asserts that the rich (path #1) and the poor (path #2) work separately to deforest the tropics. Partisans of this approach frequently argue about the amount of the world's tropical deforestation that can be attributed to the rich or the poor. Some argue that the rich do most of the damage (Painter and Partridge 1989; Fifer 1982). Others point to the poor as the primary agents of destruction (Myers 1984). The growth coalition—lead institution model asserts that rich and poor usually work in combination to clear

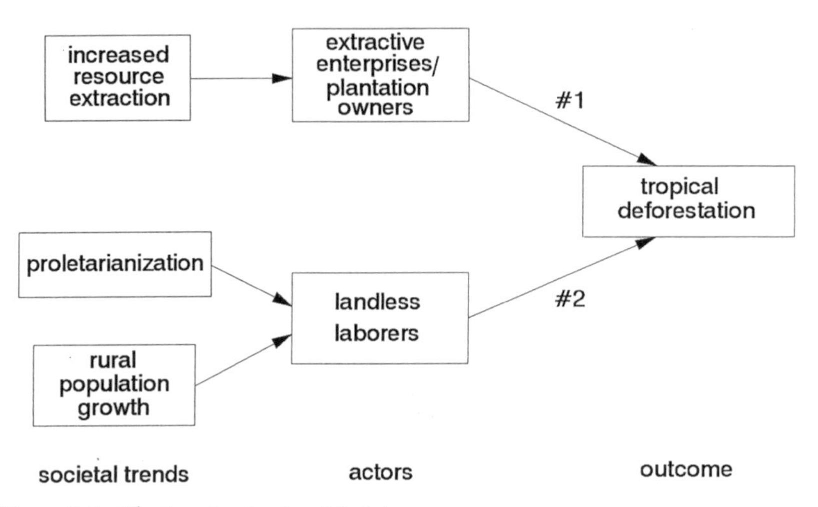

Figure 2.1. The Immiserization Model

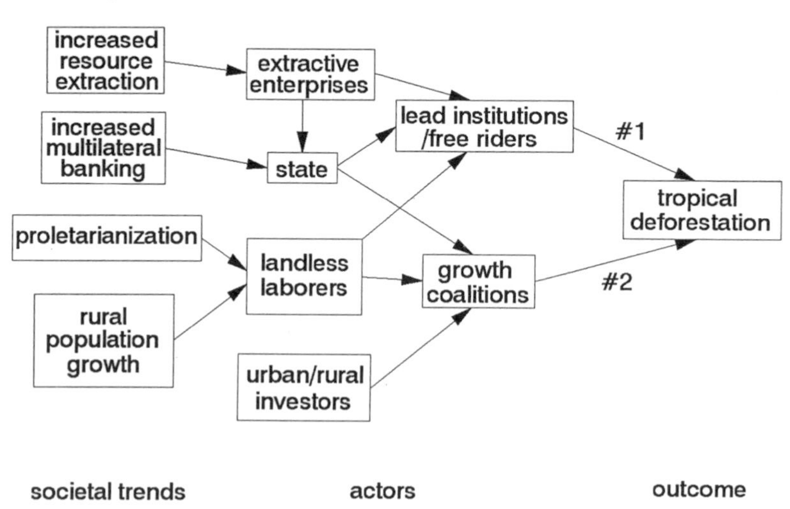

Figure 2.2. The Growth Coalition--Lead Institution Model

land. They interact in a patterned way. The rich take the lead and the poor usually follow.

The other difference between the two models concerns the roles of population growth and proletarianization in tropical deforestation. In the immiserization argument they are primary causes. In the alternative argument they provide free riders who follow roads and recruits for coalitions. By themselves, landless laborers and shifting cultivators account for little of the deforestation that occurs in large forests. Because the last assertion contradicts much of the received wisdom about deforestation, it requires close scrutiny with a wide variety of data.

Global surveys of tropical deforestation provide some indirect support for the argument presented here. F.A.O.'s most recent global assessment of tropical deforestation claims that higher rates of deforestation characterize the more fragmented forests (Dembner 1991:43). This focus on the degree of fragmentation echoes the distinction made here between islands and blocks of forest. Historical trends in deforestation provide indirect support for the growth coalition—lead institution model. Although good baseline measures of deforestation during the 1940s and 1950s do not exist; most observers would agree that rates of deforestation have increased sharply since World War II. Similar trends in two of the primary causes of the deforestation outlined in figure 2.2, rural population growth and the availability of capital, make the argument seem more plausible. Because the dramatic increase in third world populations only began after World War II, population growth has probably exerted growing pressure on forest resources since the 1950s. About the same time the creation of international development banks, coupled with an intensified search for raw materials like oil, made capital more available for regional development on all three continents.

The historical coincidence of increased rates of population growth, increased flows of capital, and rising rates of deforestation suggests but does not prove that capital and population growth work together to elevate rates of deforestation. Another aspect of the global pattern suggests that capital and labor work in combination. Rates of deforestation are much higher for secondary forests (2.06% per annum) than they are for primary forests (.27% per annum) because, presumably, lead institutions have already built access roads into areas of secondary growth, which in turn attracts free-riding peasant colonists into these areas. A third aspect of the global pattern clarifies the extent to which the combination of capital and labor causes tropical deforestation. Among countries that have small rain forests only population growth predicted the rate of deforestation during the 1970s. The forests were so small in these places that significant amounts of deforestation could have occurred without additional expenditures on infrastructure. The forested area shrinks when nearby farmers increase the size of their fields in order to provide for growing families. In countries with large blocks of rain forest during the 1970s the wealth of a country along with its population growth rate predicted the rate of deforestation. Among countries with large forests only those that had considerable wealth and rapid population growth generated high rates of deforestation because only these countries had the capital to build penetration roads into forested regions and the surplus labor to clear the land (Rudel 1989a).

While the global studies provide indirect support for the growth coalition—lead institution hypothesis, only detailed ethnographic studies can answer essential questions about the hypothesis. Descriptions of how growth coalitions accelerated deforestation in particular places would make the argument more plausible. Accounts of how economic booms and busts affected the formation of land clearing coalitions would clarify the impact of macroeconomic factors on processes of deforestation. Finally, a comparative case study could address questions about indigenous peoples. The argument assumes that rural inhabitants have a strong market orientation despite the presence of indigenous peoples throughout the tropics who have only partial commitments to participation in market economies. If the case studies demonstrate a close association between growth coalitions and deforestation among indigenous peoples as well as peasants, the explanatory potential of the argument increases.

To answer these questions, we carried out a case study of deforestation in the upper reaches of the Amazon basin in southeastern Ecuador. Beginning in the 1920s, peasant migrants and investors from the Andean highlands sought lands in the rain forests to the east. The seventy-year history of their efforts provides evidence about the role of growth coalitions and lead institutions in opening up a region and the role of population growth in subsequent land clearing. Because

Ecuador experienced an oil induced economic boom from 1972 to 1982 followed by a sharp recession, macroeconomic conditions varied dramatically during the period under study. By tracing through the impact of these changes on road building in the area, we should be able to gauge the effects that changes in the larger economy have had on deforestation rates. Because the region under study contains a large indigenous group, the Shuar (Jivaro), as well as mestizo colonists, the case study should also illustrate how ethnicity can affect patterns of land clearing. By illustrating how coalitions, population growth, macroeconomic fluctuations, and ethnicity influence land clearing, the case study should illuminate both the problems and the potential of the growth coalition argument about deforestation. The presentation of materials begins in the following chapter with descriptions of Ecuador's political economy and its rain forests.

3
The Ecuadorian Amazon: Land, People, and Institutions

■■

Why choose Ecuador for a case study of tropical deforestation? Methodological considerations played an important role in the choice of Ecuador as the site for the study. The deforestation problem takes a representative form in Ecuador. With 142,000 km2 of rain forest Ecuador is one of nineteen countries that have large blocks of humid, tropical forest (Office of Technology Assessment 1984). The destruction of its rain forests raises all of the now familiar questions about global climate change and declining biodiversity. Although its land area is only slightly larger than the state of Colorado, Ecuador has more plant species than all of North America and more animal species than the continental United States.[1] Because the vast majority of these plants and animals live in Ecuador's rain forests, the continued destruction of these forests could cause massive species extinctions.[2]

Recent studies of global deforestation usually group Ecuador with four other countries, Brazil, Colombia, Indonesia, and Malaysia, as places with large forests and rapid rates of deforestation (Ledec 1985:183). Estimates of the rate of tropical deforestation vary from 100,000 to 340,000 hectares per year with the most precise studies yielding the highest estimates.[3] According to the most recent study, Ecuador has the second highest rate of deforestation, 2.3 percent per year, in Latin America (World Resources Institute 1990:36). With its extensive forests and rapid rates of deforestation, Ecuador provided an appropriate setting for illustrating how the causes of deforestation change over time as large forests decline in size. My long familiarity with Ecuador's Amazon region made it easier to secure the necessary permits and identify field sites in Ecuador than it would have been elsewhere.[4]

While Ecuador has rain forests to the west of the Andes, most of the forests and almost half of the country's land area, 49.8 percent, lie east of the Andes in the upper reaches of the Amazon basin (Ministerio de Agricultura 1977:4).[5] Over the past thirty years, almost all of the deforestation has occurred in three locales: on the northern coastal plain, on the eastern slope of the Andes in the province of Morona Santiago, and in the northern Oriente in areas opened up by the oil companies around Lago Agrio.[6] Our research focused on Morona Santiago (see figure 3.1).[7] Situated in the southern Oriente, Morona Santiago covers 22,229 km2, an area slightly larger than the state of New Jersey, and had an estimated population of 95,000 people in 1990.[8] Approximately one-third of the inhabitants belong to a lowland indigenous group, the Shuar.[9]

Special circumstances obscure an underlying similarity in land clearing processes in Ecuador's three rain-forest regions. Timber companies played a significant role in clearing land along Ecuador's northern coast, but smallholders working in corridors along highways have cleared the most land (Barral 1979). Oil companies triggered rapid deforestation in the northern Oriente when they constructed roads to service their wells and pipelines, but colonists working small tracts of land along the roads have cleared the most land. Foreign-owned African palm plantations, despite their notoriety, have cleared relatively little land in the northeast (Carrion and Cuvi 1985). Smallholders have cleared almost all of the land in the southern Oriente. The influx of smallholders has created similar agrarian structures in the three regions. Smallholders predominate, and an intense competition

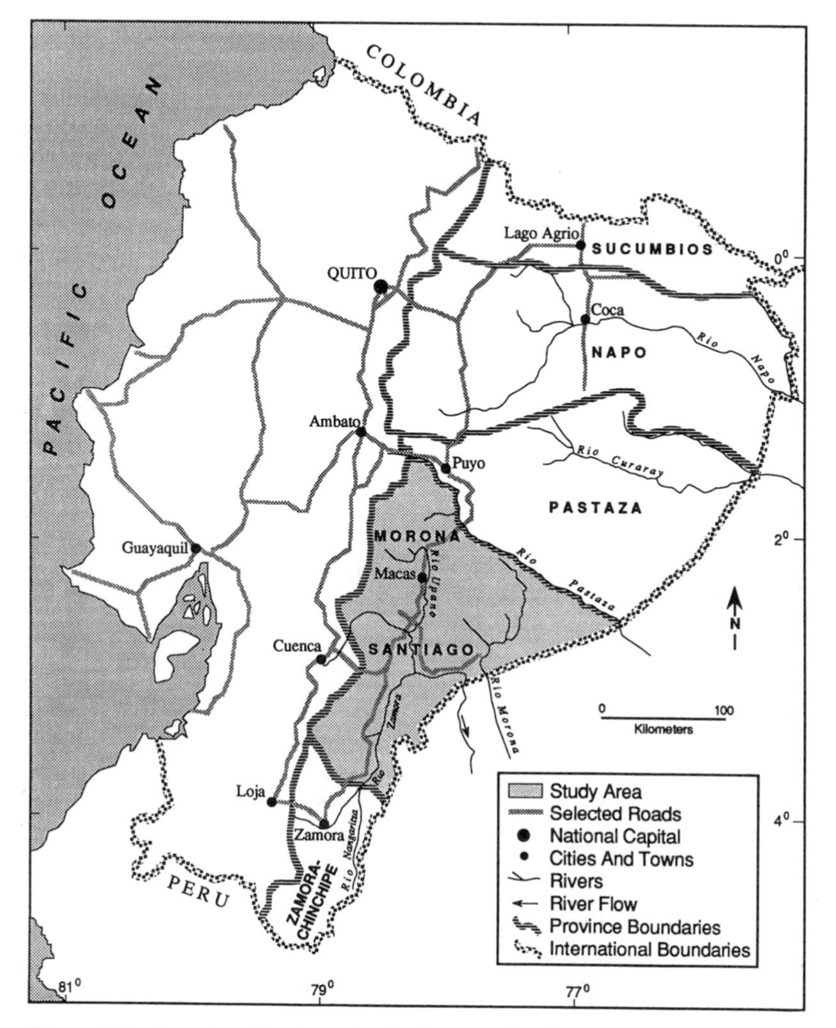

Figure 3.1. Ecuador: Provinces in the Amazon Region

for land between colonists and indigenous peoples characterizes local politics (Chiriboga, Landin, and Borja 1989:23–24). The same pattern of smallholder predominance and colonist–Amerindian conflicts characterizes other colonization areas such as Rondonia in Brazil and Caqueta in Colombia. In this sense land clearing in Morona Santiago proceeds in a familiar political-economic context.

THE RAIN FOREST AND ITS PEOPLE

The eastern slope of the Andes and spurs from the Andean chain, the Cordillera de Kutucu and the Cordillera del Condor, dominate the horizon in Morona Santiago (see figure 4.1). At high elevations narrow river valleys cut the Andean massif. The rough trails established by the first colonists wind down these valleys into the rain forests. Farther east at lower elevations, valleys with subtropical climates alternate with mountain ranges. Still farther east, on the other side of the Kutucu range, a large plain, broken by low ridges, extends to the border with Peru. The region's rivers run fast over accentuated terrain. In most places they are difficult to cross and impossible to navigate. Under these circumstances people use mule trails and roads to travel in the region.

The elevation of major population centers ranges from 1100 meters at Macas, just west of the Andean massif, to 300 meters at Morona, on the border with Peru. Mean annual temperatures vary, inversely with elevation, from 21C to 25C. It rains a lot. The residents of one place in the shadows of the Andean escarpment often say, with a smile, that "there are only two types of weather here, drizzles and deluges."[10] The westward movement of trade winds up the mountain slopes triggers orographic rainfall. Annual rainfall increases from 2000 mm in the southern portions of the province, where the spurs from the Andean chain have created a rain shadow, to 4850 mm on the province's northern border (Ferdon 1950:37, 67–68). The region does not have pronounced seasons; daily temperatures vary little from month to month, and rainfall occurs evenly throughout the year; when dry spells occur, they never last more than one or two weeks (Grubb and Whitmore 1966:305–7). Almost everyone acknowledges that the climate has changed in recent years, becoming noticeably drier, but the absence of detailed meteorological records makes it impossible to estimate the amount of change.

In a pattern common throughout the tropical world (Lele and Stone 1989:21), the most extensive settlement in Morona Santiago has occurred in areas with infertile soils. The predominant soils in the settled, western valleys are Hydrandepts with significant amounts of volcanic ash. They are deep, relatively loose, and waterlogged. These soils are also acidic, saturated with aluminum, and deficient in potassium, all of which limit their agricultural potential. Small patches of soil with more organic material occur in poorly drained swampy sites at the base of hills. The ministry of agriculture recommends that these areas remain forested. West of the Kutucu mountains the sparsely settled plain contains bands of fertile soils in alluvial deposits along major rivers (Ministerio de Agricultura 1980a).

Lowland rain forest predominates in areas below 1200 to 1500 meters in elevation (Richards 1952:363). In life zone terminology the western portions of the study area extend into subtropical and premontane environments, but most of the region contains either tropical moist forests, tropical wet forests, or tropical rain forests (Holdridge 1967). The forests contain three strata of vegetation and trees which reach 80 to 120 feet in elevation (Grubb, Pennington, and Whitmore 1963:596). Climbers and buttresses adorn many of the trees. Except in places where thick groves of bamboo (*Bambusa*) or tree falls permit light to get through, the forest floor is dark; the vegetation is sparse, and shades of brown and green color everything. The canopy is so thick that oncoming showers announce their arrival with the low humming sound of rain hitting leaves several kilometers away. Above 1200 meters the trees decline in size, and the canopy opens up in some places.

For at least five hundred years the Shuar, who belong to the Jivaroan group of peoples, have inhabited the rain forests of Morona Santiago. In addition to hunting and gathering, the Shuar practice a slash-mulch polyculture (Hiroaka and Yamamoto 1980:438). The Shuar must clear the forest when it is wet because dry seasons appropriate for clearing and burning do not occur. Given the moisture in the forest, the Shuar cannot burn the felled vegetation, so they leave it to rot, providing a mulch for cultivated plants. Although a typical Shuar garden might contain between ten and twenty cultivars, manioc

(*Manihot esculenta*) dominates the gardens. Manioc plants make up more than 80 percent of the Jivaroan gardens in northern Peru (Boster 1983). Weed infestation forces the Shuar to abandon old gardens and establish new ones every three to five years. Growing scarcities of firewood force the Shuar to move to new house sites every six or seven years (Harner 1973:44–48).

Until the mid-twentieth century Shuar families lived in a dispersed pattern with houses located up to one-half day's walk from each other (Karsten 1935). Trade networks extending over hundreds of kilometers linked friendly households to one another. Conflicts between households defined social relations outside the family. Households either clashed or coalesced with one another in recurrent conflicts (Hendricks 1986:4). One enduring pattern of conflict pitted the Shuar living west of the Rio Macuma against the linguistically and culturally similar Achuar living east of the Macuma (see figure 4.2).

The Shuar acquired a bellicose reputation as early as 1527 when the Incas tried unsuccessfully to subdue them (Harner 1973:17). Briefly reduced to slavery by Spanish conquistadors, the Shuar rebelled in 1599, sacked the mining cities of Sevilla de Oro and Logrono de los Caballeros, and killed all of the male inhabitants (Ampudia 1987). High levels of conflict and homicide among the Shuar during the past two centuries have reinforced their reputation for aggressive behavior. In the early twentieth century internecine feuding among the Shuar contributed to so many homicides, followed by the taking of heads, that their population began to decline (Karsten 1935:2). The frequency of these disputes convinced early ethnographers that the Shuar were "an individualistic people, intensely jealous of their freedom and unwilling to be subservient to authority, even among themselves (Harner 1973:1)." With this disposition the Shuar do not hesitate to complain when slighted, and the complaints cause conflicts. According to another ethnographer of Jivaroan peoples,

> there is a strong disposition towards hostility in all social relations, and people quickly suspect and mistrust, which provokes a large number of disputes and tensions at all levels of society: within the family, within the community, and between communities (Seymour-Smith 1988:168).

This hostility includes an intense suspicion of outsiders which has slowed but not stopped attempts by the church and colonists to dominate the Shuar. Several Catholic missionary groups, the Franciscans and the Salesians, withdrew from the Oriente after unsuccessful attempts to convert the Shuar in the early twentieth century, but by mid-century both groups had resumed their missionary work (Bottasso 1982:20–21; Salazar 1981:593). Throughout the twentieth century fear of armed conflict has prevented some colonists from occupying Shuar lands, but other colonists have pressed forward, and the ensuing struggles have led to shootings and deaths.[11]

During the 1940s and 1950s the missionaries and the national police managed to suppress most of the feuding among the Shuar. The decline in homicides and the introduction of modern medicines like antibiotics reduced death rates among the Shuar, and their population began to increase (Harner 1973:211). By the mid-1980s approximately 35,000 Shuar and 2,400 Achuar lived in Morona Santiago.[12]

THE STATE AND THE ORIENTE

For centuries the eastern slope of the Andes presented a formidable physical obstacle to any kind of organized activity by states, corporations, or missionary orders. While the placid, downstream stretches of the Amazon's tributaries made it relatively easy to move people and goods in the lower reaches of the Amazon basin, no one could navigate the rivers rushing out of the Andes, so everyone had to travel by foot in the upper part of the basin. With the advantages of river travel Peruvians based in Iquitos controlled commerce in large parts of the Ecuadorian Oriente during the nineteenth century. In the early twentieth century Peruvian traders traveled up the tributaries of the Amazon to establish networks of rubber tappers in Ecuadorian territory. Anxious about the Peruvian presence, Ecuadorian officials tried to strengthen their claims to the Oriente by creating a program of military colonization in which veterans of military service would form communities along the nation's Amazon borders.[13] Chronic fiscal problems made it impossible to fund the program, so the Amazon region remained only loosely integrated into the nation's political and economic life.

The problems of national integration that worried Ecuadorian elites extended far beyond the question of how to incorporate the Oriente into the nation's affairs.[14] One overarching question dominated political discussions in the early twentieth century. Could one country unite people from different highland provinces who rarely saw one another and frequently spoke different languages (Quichua and Spanish)? The absence of contact between the peoples of different highland provinces generated an intense regionalism that had a profound impact on Ecuadorian efforts to incorporate the Amazon region into the national economy.

Political Regionalism and the Oriente

Without precious metals or fertile lands close to markets, Ecuador attracted little foreign capital during the nineteenth century, and it's economy stagnated (Rodriguez 1985). Because little long distance trading occurred across the mountains, the topography of the Andes mountains defined the boundaries of local economies. The Andes run north to south in two parallel ranges, the Cordillera Occidental and the Cordillera Oriental, with a central valley between the ranges. Transverse ridges several thousand feet high, known as *nudos*, unite the two *cordilleras* at intervals, so the central valley is really a series of basins separated by ridges (Hermessen 1917:434). Seasonal rains combined with the rugged topography made the trails between the basins impassible for months at a time, so nineteenth-century Ecuadorians traded little and consumed much of what they produced. The isolation of the populations in the different basins continued into the twentieth century. Loja, for example, had no road linking it with Sierra provinces to the north until the 1940s.

The geographic isolation fostered intense regional loyalties that affected the organization of Ecuadorian politics. Each basin became a separate province, and provincial representatives in the central government concerned themselves primarily with their basin's interests. Political parties defined themselves by regional interests as well as by ideology. Separate liberal parties represented, for example, the interests of liberals in the Sierra and liberals on the Coast. The pattern of public works expenditures provides further evidence of the strength of regional interests. To satisfy the demands of regional elites, the ministry of public works spread its expenditures out over a large number of projects, each in a different province (Rodriguez 1985:190). With each project receiving only a small allocation of funds each year, construction proceeded slowly, and projects remained unfinished for years. Collectively, the regional coalitions created a political order which earmarked separate portions of the federal budget for each province. This practice reduced the ability of the central government to undertake new initiatives of any kind (Thoumi 1990).

When provincial elites thought about the Oriente in the early twentieth century, they extended the regional divisions of the Sierra into the Oriente. Rather than conceiving of the Amazon region's abundant resources as a solution to the nation's rural poverty, people regarded the portion of the Amazon region immediately to the east of their highland province as a solution to their province's rural poverty. Elites in the city of Cuenca in Azuay province referred to the lowlands to their east as the "Oriente Azuayo" (Salazar 1986:51). For them the lowlands represented a potential colony whose development would stimulate commercial activity in their Andean homeland. To open up their portion of the Oriente, each provincial government in the Sierra formed a commission to promote the improvement of trails and, later, the construction of roads into the Amazon basin (Jaramillo Alvarado 1936:328–329).

To appease sectional interests, the central government made small annual appropriations for each province's penetration road. In 1962, for example, the government appropriated funds for the construction of ten different roads from the Sierra to the Oriente. To complete each of these roads, the contractors had to build 70 to 90 kilometers of road over extremely mountainous terrain. Under these conditions the roads inched forward. At 1962 levels of funding the roads would have taken an average of eighty years to complete![15] The appropriated funds were not only insufficient to make significant progress towards the completion of a road, they were often available only intermittently. The history of the Sigsig–Gualaquiza road illustrates the problem (see figure 4.1). Begun during the mid-1950s, the road lost its funding in the early 1970s, had its funding restored during the late 1970s, only to lose its funding again during the early 1980s. In thirty-five years of construction the road has advanced only twenty-eight kilometers![16] By spreading the funding out over so many projects and providing it intermittently, the central government in effect decided against a national effort to construct a single penetration road and, in so doing, delayed the opening of the Amazon region to outside economic interests (Christ and Nissly 1973:90).

Foreign Colonization

To supplement national efforts to develop the Amazon region, government officials began to promote foreign colonization in the early twentieth century. Between 1910 and 1940 the Ecuadorian government signed contracts with a French company, an Italian missionary order, and agencies in Austria, Czechoslovakia, and Spain to populate the Oriente with emigrants from Europe (Perez Guerrero 1954:145; Bottasso 1982:109–110). In a typical agreement the government promised to build a railroad from the Sierra into the central Oriente and agreed to give each colonist sponsored by a French company, L'Explotation de Concessiones Ecuatoriennes, two hundred hectares of land in the Oriente. In return the company agreed to settle four thousand young Europeans in the region.[17] The agreement collapsed when the government made no efforts to construct the railroad. After World War II a conservative president, Camilo Ponce Enriquez, offered to settle large numbers of European refugees on forested lands in the Oriente. As late as 1978 conservative journalists urged the government to settle displaced white settlers from southern Africa in the Oriente (Bottasso 1982:110).

The argument for foreign colonization had several sources. First, high mortality rates reduced population growth rates during the first half of the twentieth century, so Ecuadorian elites did not perceive a problem of excess population which the "empty" lands of the Oriente might solve. Second, many highland peasants could not move; they were tied to the land through the *huasipungaje*, a labor for land arrangement that made it impossible for them to leave the hacienda on which they lived (Perez Guerrero 1954:133). Third, foreign colonization would in most instances cost the state very little because many of the colonists from Europe and Japan would come to Latin America with subsidies from governments in their countries of origin (Crespo 1961:203–242). Finally, foreign colonization attracted many influential Ecuadorians for racial reasons. If successful, foreign colonization would bring a stream of white immigrants who would reinforce the self-anointed elites of Spanish ancestry who both exploited and scorned the large Amerindian population in the highlands.

While the foreign colonization schemes invariably failed to bring significant numbers of new settlers to the Amazon region, foreigners still played an important role in the initial settlement of the region. Catholic missionaries, almost all of them foreign born, established most of the first settlements in the southern Oriente. Italians predominated among the Salesian missionaries in Morona Santiago to such an extent that their superior between 1921 and 1961 often communicated with them in Italian rather than Spanish (Bottasso 1982:68). While some foreigners opened the region up in order to save souls, others did so for more earthly reasons. Funds from foreign oil companies and banks proved important in the construction of three of the first four penetration roads to reach the Oriente from the Sierra. An American oil exploration company began the Ambato–Puyo road during the 1920s, and the Ecuadorian government completed the road during the early 1940s only after Shell Oil, an European company, agreed to build six airfields and drill five exploratory oil wells in the central Oriente (Tschopp 1953:2305). Thirty years later an American company, Texaco, constructed the Quito–Lago Agrio road. During the late 1960s and early 1970s the Interamerican Development Bank provided the financing for the construction of the Cuenca–Macas road into Morona Santiago (see figure 4.1).

The salience of foreigners in these early efforts at settlement reflects the weakness of the Ecuadorian state. Even though the Ecuadorians lost a large expanse of the Amazon basin to Peru in the disastrous war of 1941, the central government was too starved for funds and too preoccupied with interregional conflicts after the war to launch a concerted program to develop what remained of the Amazon region. For example, when the Shell Oil Company failed in their initial efforts to find oil in the late 1940s, President Galo Plaza declared that the supposed agricultural riches of the Oriente were "a myth" and that the government should focus its limited resources on the agricultural development of the coastal plain to the west.[18] Under these circumstances the government was only too willing to cede practical control over parts of the Oriente to missionaries and to rely on foreign funds for regional development initiatives in the Oriente. With the withdrawal of the oil companies, only the poorly funded missionaries provided any impetus for regional development, so deforestation proceeded slowly.

The Modern Oriente: Agencies, Companies, and Migrants

The increase in government activity during the past thirty years has changed the identity of the important institutions in the Oriente, but with one exception it has not altered the calculus that landowners use in deciding whether or not to clear land. The exception involves land reform laws. The first reform, enacted in 1964, reorganized bureaucrats in government and peasants on Sierra haciendas, but it had little effect on land use anywhere in the country. The second reform, enacted in 1973, changed patterns of land use in both tropical and temperate regions.

The Cuban revolution crystallized popular discontent with Latin America's landholding system during the early 1960s and prompted conservative organizations like USAID to advocate land reform as a preemptory reform in order to prevent popular revolution (Grindle 1985; Feder 1971). In Ecuador the pressures for reform came from internal as well as external sources. Leftist politicians advocated reform, but, surprisingly, so did a number of large landowners in the Sierra. The advocacy of reform by large landowners reflected changes in peasant-landlord relations in the Sierra. With the spread of modern medical technologies, peasant populations on the large haciendas had begun to grow at the same time that mechanization had reduced the landlords' need for labor. Under these circumstances modernizing landlords saw in a land reform an opportunity to rid themselves of a growing peasant population to whom they had traditional obligations of providing land for gardens and pasture (Barsky 1984:55–166).

The 1964 reform abolished the traditional *huasipungo* arrangement, and created an agency, the Instituto Ecuatoriana de Reforma Agraria y Colonizacion (IERAC), to carry out the land reform and initiate colonization projects. The 1973 reform tried to stimulate production on large, underutilized landholdings by requiring that landowners cultivate at least 50 percent of their land if they wanted to avoid expropriation by IERAC. Another law, enacted in 1978, created a new agency, Instituto Nacional de Colonizacion de la Region

Amazonica Ecuatoriana (INCRAE), to plan and administer colonization projects in the Oriente (Hicks et al. 1990:35).

The first reform had little impact on deforestation rates because most colonists came from the southern highlands, an area unaffected by the reform. The second reform, requiring that farmers cultivate at least 50 percent of their land, had a major impact. The 50 percent provision, coupled with colonist pressures on Amerindian lands, convinced the Shuar and the lowland Quichua of the central Oriente that they would have to clear large portions of their land in order to retain control over it. Accordingly, both the Shuar and the lowland Quichua cleared extensive tracts of land for pasture and began to acquire cattle during the mid-1970s, right after the enactment of the second reform (Federacion de Centros Shuar 1976; MacDonald 1984).

Peasants have continued to migrate to the rain-forest regions throughout the twentieth century despite large changes in other migratory streams in Ecuador. The 1960s and 1970s saw a sharp increase in geographical mobility among Ecuadorians. While only 13 percent of all Ecuadorians moved between 1950 and 1962, 28 percent of the population moved between 1974 and 1982. Most of the migrants moved from small villages to large cities. Between 1950 and 1982 the proportion of the nation's population living in urban places of 50,000 or more persons increased from 14.6 percent to 38.3 percent (CONADE 1987:225). The Amazon region received a relatively small stream of migrants throughout this period. By 1990 it had approximately 350,000 inhabitants, a little less than 4 percent of the nation's population. Almost half of these residents had moved to the Oriente from other parts of the country (Hicks et al. 1990:1). The numbers of inhabitants in the Amazon region remained small, but they used land liberally. While the proportion of the nation's population living in the Amazon increased from 1.9 percent to 3.3 percent between 1962 and 1982, the Amazon's portion of the nation's cultivated land jumped from 0.9 percent to 7.6 percent (CONADE 1987:207).

Land distribution in the Amazon basin reflects recent settlement processes. Smallholders with tracts of twenty to one hundred hectares own most of the land.[19] In theory the central government tries to assist smallholders through integrated rural development programs (IRD) that build infrastructure, deliver extension services, and provide credit to small farmers. In fact the government has carried out only one IRD in the Oriente despite the concentration of smallholders in the region (Chiriboga, Landin, and Borja 1989:7). The location of the one IRD, in the Nangaritza Valley, near a large gold mine and a disputed section of the border with Peru, suggests that it owes its existence to extraordinary geopolitical considerations.

Official inattention to smallholders in the Amazon region extends to other agencies. INCRAE, the agency officially responsible for colonization in the Oriente, has not initiated any major colonization projects during its twelve year existence. The periodic press releases of IERAC, the agrarian reform and colonization agency, suggest that it has delivered thousands of hectares of land to new colonists. In fact it does no more than issue land titles to spontaneous colonists who settled their lands without any help from IERAC. The headlines serve as a political palliative in the nation's capital, suggesting a continued commitment to the ideals of agrarian reform and colonization. In the early 1970s the editorial writers at a newspaper in one Sierra city noted, with some disgust, the rhetorical uses of the Amazon region in Ecuadorian politics.

> The Oriente gives us a pretext for carrying out publicity campaigns, electoral campaigns, and supposedly great programs of action and improvement. Despite its tremendous wealth and thanks to the above mentioned game, the Oriente and its incorporation into the active life of the country continues to be one more formula, slogan, and hope like many others. For years our political adventurers and financiers have manipulated us with these formulas and slogans for their own personal political benefit.[20]

This pattern of rhetorical commitment and real neglect of Oriente agriculture reflects the ability of well established sectional interests to earmark funds for themselves and fiscally starve new, less powerful constituencies like smallholders in the Oriente. In this instance agroexporters from the Coast shaped agricultural policy and insured that smallholders in the Amazon received little attention (Lawson 1988). Government officials departed from this policy only once, in the mid-1970s when tax receipts quadrupled after the country began to export oil from the northern Oriente. Every government agency participated in the bonanza. The Banco de Fomento received a major infusion of capital and for the first time offered credit to small farmers at subsidized rates; the regional development agency for Morona Santiago, the Centro de Reconversion Economica de Azuay, Canar, and Morona Santiago (CREA), received full funding for its colonization program. In 1975 IERAC saw its budget triple in size over its 1974 levels (Martz 1987:121–122, 253). This bonanza quickened the pace of settlement and deforestation between 1975 and 1980. The debt crisis of the early 1980s ended the federal government's largesse, and peasant colonists in the Amazon had to seek assistance elsewhere.

Some of the assistance came from foreign capitalists who continued to play a vital role in the development and deforestation of the region. Foreign financed road construction attracted colonists to two areas during the late 1980s. A French oil company, Elf Aquitaine, constructed a road south from Coca (see figure 3.1) to exploit a new oil field, and colonists settled on both sides of the road.[21] The Brazilian government loaned the Ecuadorian government funds to employ a Brazilian company, Andrade Gutierrez, to build a penetration road into Morona (see figure 4.1).[22] Colonists from western Morona Santiago have staked out claims and begun clearing land along the new road.[23]

The extensive road building over the past thirty years has changed the ways colonists choose land for clearing, but the importance of agricultural expansion to smallholders has not changed. The government provides little institutional support for efforts to intensify agriculture, so smallholders see expansion of the area under cultivation as the only way to increase their income. The government's continuing neglect of intensified agriculture on small farms has promoted an extensive pattern of agriculture that entails clearing large tracts of forest. In its neglect of smallholder agriculture the Ecuadorian government is in no way exceptional; governments in most other developing countries have also neglected peasant agriculture in recent years (Johnston and Clark 1982).

While this contextual analysis has identified the larger historical conditions that have slowed or speeded deforestation, it has not identified the proximate conditions that from year to year and place to place either enable or prevent land clearing. The growth coalition hypothesis provides the missing link in this argument. It outlines the sequence of events, actions, and changed conditions that open up regions for settlement and accelerate rates of deforestation. To assess its empirical validity, we will take a close look at patterns of development and deforestation in Morona Santiago between 1920 and 1990.

4

Regional Development and Deforestation in Morona Santiago

THE SENDING REGION: THE ANDEAN HIGHLANDS OF SOUTHERN ECUADOR

Early in the twentieth century peasants in the southern highlands (Azuay and Loja provinces) began to see colonization in the Amazon rain forests as the way to a better future. The rural poor in the southern Sierra differed from their counterparts in the northern Sierra in ways that made migration to the eastern rain forests a more realistic option for them. In the late 1950s more than 40 percent of the peasants in the northern highlands worked as peons on large haciendas and could not move about freely. Only 10 percent of the peasants in the southern highlands found themselves in a similar situation. The remaining 90 percent lived on small, independently owned plots of land from which they could move in search of a livelihood (CIDA 1965:78). The southerners had good reason to want to move. Erosion had made much of the land in Azuay unworkable, and a semi-arid climate (600 to 800 mm of rainfall per year) insured frequent crop failures, so agriculture did not provide a secure livelihood. As one migrant put it, "we planted hope, but we harvested dust and drought."[1]

To meet their subsistence needs, southern peasants looked for economic opportunities in a wide variety of places. During the 1930s 8,000 to 10,000 men spent several months of each year panning for gold in streams on the eastern slope of the Andes (Reyes and Teran 1939). At home men and women found steady, but poorly paid work as artisans in the Panama hat industry. People wove hats while carrying on innumerable other tasks, cooking, nursing a baby, herding sheep, or walking to market. When the international market for hats collapsed during the 1950s, the already meager incomes of the weavers declined to lower levels, and more men began to look for work outside the region. The largest number went to the coast to work as laborers on banana plantations. A smaller number began looking for new lands in the sparsely populated, subtropical valleys beyond the eastern crest of the Andes, some three to seven days' walk from their homes in the highlands.

Sparse historical records make it impossible to identify who moved to the Oriente first. Because the pioneers relied on kin for financial support until the farm in the Oriente began to produce, only families with some resources could send someone over the mountains. A study of emigration from one Azuay community, Gima, supports this conjecture. "It is not the poorest Gimenos who have land in the . . . (Oriente) but those that are fairly secure and . . . able to take the risks involved in migrating to the eastern valley" (Eckstrom 1979:36). In addition to help from kin, pioneers could in many instances count on assistance from missionaries who had recently established outposts on the edge of the rain forest.

DEVELOPMENT IN THE RAIN FOREST: COLONISTS AND THE SALESIAN MISSIONARIES

At the turn of the century Morona Santiago had one agricultural community, a group of 400 mestizos, who farmed a small area around

Macas in the upper reaches of the Upano valley.[2] The trail linking Macas with the Sierra was so steep that neither mules nor horses could use it. Without pack animals to transport goods Macas' farmers could only sell what they could carry on their backs up to highland markets in Riobamba.

Under these circumstances farmers produced mostly for subsistence. Government officials did little to bring the Oriente into the orbit of national life. They authorized missionary work because it would tighten the government's grip on the region without the expenditure of public funds. In a 1911 interview a government minister applauded "the work of the (Salesian) missionaries, not only because they would convert the native peoples, but, because they would work to cement our sovereignty in places which are constantly eyed by our acquisitive and active neighbors (from Peru)." The minister for the Oriente stated that he wanted "missionaries who would be an important factor in securing our national sovereignty over these regions."[3]

In 1894 the Salesian Order began to send Italian priests into the jungle to work with the Shuar. The first twenty years taxed the missionaries' spirits. Having established a mission in Gualaquiza, they found it virtually impossible to initiate any sustained contact with the Shuar (Bottasso 1982:21–22). Under these circumstances the Salesians turned their attention to the small numbers of Spanish-speaking colonists from the highlands. The Salesians assisted the colonists in a variety of ways. In 1914 Revs. Albino del Curto and J. Bonicetti helped the first colonists in the Limon-Indanza region establish a church and a village center.[4] The first two families of colonists to arrive in Mendez received foodstuffs and shelter from the mission in return for working several days a week on church lands.[5] Funds for the first schools, bridges, and trails in these communities came from Salesian sources; the government provided no funds for public works in the southern Oriente until the 1940s.[6]

For decades rough trails from the highlands limited settlement in the valleys at the base of the Andes. A Salesian missionary described one trail as "only centimeters wide, closed off from the sky by jungle growth, obstructed by rotting tree trunks at each step, transformed into river beds in some places, . . . and covered by thick layers of mud, interspersed with potholes, in other places" (Allioni 1910, quoted in ALOP-CESA 1984 2:429).

Travelers frequently waited for days on the banks of a swollen river, hoping that receding waters would allow them to ford the river or scramble across a makeshift bridge (Spruce 1908 2:102–170). Neither cattle nor mules could negotiate the steep stretches of these trails, and the infrequent improvement projects concentrated on extending the stretches horses and mules could use (Platt 1932; Jaramillo Alvarado 1964). Because colonists could not drive their cattle to market, they had little use for large herds of cattle or extensive pastures, so land clearings remained small.

The Salesians began a long effort to improve transportation in 1917 when del Curto obtained funds from the Salesian mission, the Society for Historical Studies of Cuenca, and the national government to construct a wide, hard surfaced trail from Pan to Mendez (see figure 4.1) (Brito 1938:321, quoted in Bottasso 1982:111). As one colonist put it, del Curto was "el macho" in this effort; he scrounged for the funds, surveyed the route, contracted the workers, and accompanied them to the site of the blasting and the digging.[7] When an accident took the lives of six workers during the construction of the trail in 1919, he prayed for their souls (Jaramillo Alvarado 1936:87–97; Hegen 1966). After the completion of the trail in the early 1930s, colonists could drive their cattle from the Amazon lowlands to the highland markets in fifteen days. The cattle were 'all bones' when they arrived in the highlands, but their sale provided the colonists with incomes and an incentive to clear land for pastures.[8]

The Salesian missions became poles of attraction for newly arrived colonists. The missions provided medical care for all and schooling for the colonists' children (Bottasso 1982:115). In addition the priests provided the colonists with a political counterweight to powerful residents of the Sierra who claimed but did not occupy large areas of western Morona Santiago (Jaramillo Alvarado 1964:69–70, 135–136). While most Salesians worked with the Shuar, some priests such as del Curto, dedicated themselves to pastoral work among the colonists. The Salesians established at least one mission, Cuchantza (near Mendez), to minister to the colonists (Bottasso 1982:123). The priests working out of these missions played an important role in the development of outlying communities. When migrants from the Sierra founded the community of San Juan Bosco in 1952, a priest, Rev. Luis Carollo, distributed tracts of land to the colonists. Carollo went

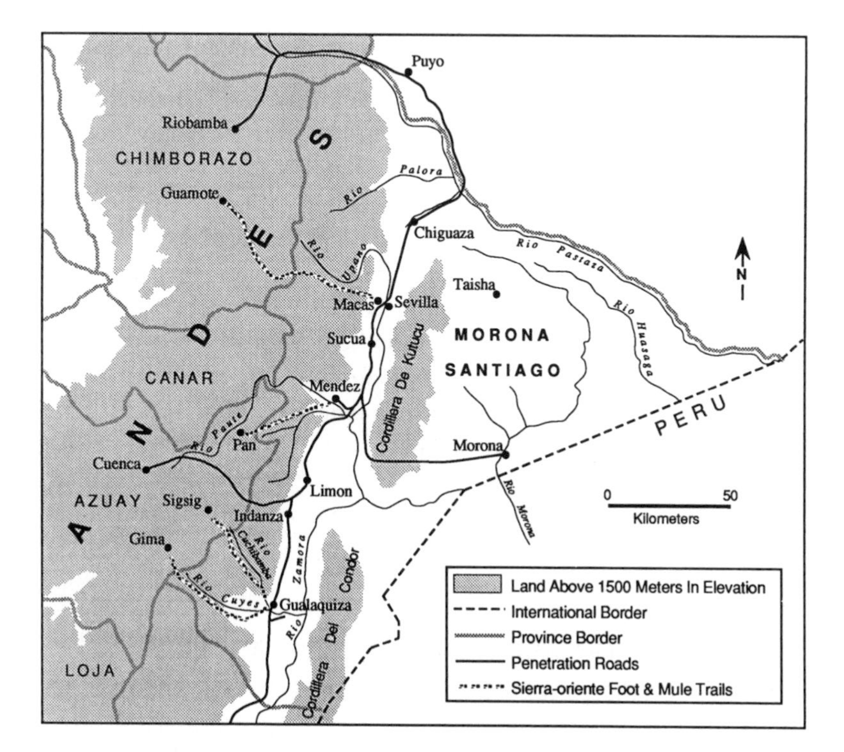

Figure 4.1. Penetration Routes in Morona Santiago

on to become the first priest in this parish; in the words of one parishioner he was "the soul of this village; he gave it form and life."[9] Even the bishop devoted himself to the care of the colonists. From the 1920s to the mid-1950s he made long, arduous trips by horse to visit priests and parishioners in outlying communities. As he put it, "the seat of my diocese is in the saddle of my horse" (Bottasso 1982:68). While the Salesians' devotion to the colonists stemmed in part from the difficulties that they encountered in their missionary work with the Shuar, the state's neglect of the colonists also strengthened the Salesians' commitment to them. The priests did so much because the state did so little.[10]

What would have happened if the Salesians had not established missions during the first half of the twentieth century? A comparison of colonization in places with and without missions should provide a preliminary answer to this question. The Salesians established missions in three of the four places in southern Morona Santiago where rivers emerge from the Andes. Migrants from the Sierra settled the fourth valley without help from the missionaries. A comparison of colonization in the adjacent Cuchibamba and Cuyes river valleys (see figure 4.1), where the Salesians did and did not, respectively, establish missions should indicate how their presence affected settlement and deforestation.

THE CUCHIBAMBA AND CUYES VALLEYS: COLONIZATION WITH AND WITHOUT THE SALESIANS

The lure of placer gold in the rivers attracted large numbers of peasants to the two valleys during the 1930s. Hostilities with the Shuar and war with Peru drove the miners back into the Sierra in the early 1940s, but they did not forget the abundant resources in the Oriente. Many of them returned to the Oriente to start farms during the next thirty years. Settlement started sooner in the valley with the missions. Colonists began to settle in the Cuchibamba valley in the 1920s, before the gold rush, but after the Salesians had reestablished their mission at Gualaquiza in the lower part of the valley. Colonists did not begin to settle in the Cuyes valley until after the gold rush of the 1930s. Between 1920 and 1950 the Salesians established missions to

Table 4.1. Migratory Trends in the Cuchibamba and Cuyes Valleys, 1962–1982

	Population, 1962	Net Migration, 1962-1974	Population, 1974	Net Migration, 1974-1982	Population, 1982
Cuchibamba Valley					
Bermejos	300	-63	354	-66	334
Chiguinda	401	-56	493	-81	454
El Rosario	545	+2	860	-330	615
Cuyes Valley					
San Miguel	—	—	211	+106	352
Amazonas	—	—	328	+31	420
Nueva Tarqui	305	+333	788	+12	995

SOURCES FOR THE DATA: CEDIG, 1985; Anuarios Estadisticas, 1965-1982.

NOTE: San Miguel and Amazonas did not become parishes with separate population counts until after 1962. We calculated net migration rates for each parish by using data from two sources: (1) annual reports of births and deaths in each parish, and (2) census data on population changes in 1962, 1974, and 1982. The reports of births and deaths made it possible to calculate natural increase in a parish for each period. By subtracting natural increase from population change for a period, we obtained net migration. For further details on these calculations, see Rudel and Richards, 1990.

serve colonists at two intermediate points along the trail up the Cuchibamba valley into the highlands. An influx of migrants swelled the populations of three settlements along the trail, Chiguinda, Bermejos, and Rosario, to the point where they became officially recognized parishes during the 1940s. Amazonas and San Miguel in the Cuyes valley did not become parishes until the 1960s (ALOP-CESA 1984:430). Colonists had cleared all of the arable land in the Cuchibamba parishes by the early 1960s.[11] Not surprisingly, young people began to leave the Cuchibamba parishes of Chiguinda, Bermejos, and Rosario in large numbers during the 1960s when the last of the unclaimed forest land disappeared (see table 4.1).[12] Colonist families in the Cuyes valley could find unoccupied forest land near their farms until the late 1970s (Eckstrom 1979:94, 111, 146). As the data in table 4.1 suggest, the Cuyes valley communities retained their young at higher rates during the 1960s and 1970s than did the Cuchibamba parishes. By the early 1980s colonists in the Cuyes valley could no longer find unoccupied land, and outmigration began to increase.[13] These differences in the historical demography of the two valleys suggests that the Salesians' activities made deforestation occur twenty years sooner in Cuchibamba valley than it would have occurred otherwise.

The Cuyes valley never had a growth pole, so settlers did not travel long distances to reach the valley; they came from Gima, a village just over the crest of the Andes from the headwaters of the Cuyes River. Neighbors in Gima became neighbors in Amazonas, and the cooperative arrangements between and within families facilitated the clearing of land and the construction of homes. The pace of deforestation in the Cuyes valley followed the rhythm of family expansion. When the sons became old enough to start families, they carved farms out of the forests on hillsides not too far from their fathers' farms (Eckstrom 1979).

Similar processes occurred in the Cuchibamba valley. Migrants organized along lines of kinship, and natural increase influenced the pace of deforestation, but the missionaries triggered the process. Their missions became growth poles which attracted a steady stream of potential colonists. The increased traffic up and down the valley acquainted travelers with the agricultural potential of the area and encouraged the settlement of intermediate points. The missions' schools and medical services made settlement in these places more attractive. The Cuyes valley experience suggests that settlement and deforestation would have eventually begun without the missionaries' presence, but their work accelerated the process.

CREA AND REGIONAL DEVELOPMENT

In the mid-1960s a regional development agency replaced the Salesians as the lead institution in the development of Morona Santiago. CREA (Centro de Reconversion Economica de Azuay, Canar, and Morona Santiago), a regional development agency headquartered in the highland city of Cuenca, launched an ambitious program of road construction in Morona Santiago in 1965. A United Nations evaluation of CREA during the 1970s concluded that the agency "exists and functions essentially as a promotional organization for the city of Cuenca and its hinterland" (UNOTC 1974:66, quoted in Armstrong and McGee 1985:167). With a directorate composed of influential city residents, CREA is an important component in an evolving set of growth coalitions centered in Cuenca. It provides the infrastructure that facilitates private investment schemes. Colonization represents this type of investment, so CREA got heavily involved in colonization projects during the 1960s and 1970s. They made up 45 percent of CREA's annual budget during the early 1970s (Warwick and McGee 1985:165). By developing Cuenca's agricultural hinterland, the projects strengthened Cuenca's position as a distribution center which in turn enhanced the financial position of the families from which the agency drew its directors. To this end, CREA began to build a penetration road from Azuay into western Morona Santiago in 1965. The agency encouraged colonization both directly and indirectly. It provided food and technical assistance to colonists on lands near the new roads, and the roads themselves spurred other colonists to settle in the region.

Some of the new settlements survived and sparked extensive land clearing while others failed and left the forests intact. A comparison of successful and unsuccessful colonization schemes should clarify the circumstances which sparked the recent wave of deforestation in Morona Santiago. Success occurs when the colonists remain in an area long enough to clear large tracts of land; failure occurs when the settlers abandon a site without clearing significant amounts of forest.

La Union de Calablas—A Success. In the late 1960s, fifty years after the original settlement of Indanza, several wealthy descendants of the first settlers embarked on another colonization effort. The sons of the original colonists, from the Lopes family, claimed about 500 hectares of land in an area known as La Union de Calablas, some thirty kilometers south of Indanza (see figure 4.2). At the same time population growth in Indanza had increased the number of families trying to eke out an existence on one and two hectare plots of land. In the mid-1970s CREA began to build a road from Indanza to Gualaquiza that would pass through La Union. At this point the Lopes began to subdivide their undeveloped landholdings into fifteen and twenty hectare tracts of land which they sold to landless or near landless settlers from Indanza.

To speed the sale of lands, the Lopes sponsored the creation of an urban center in the area. They set aside a 3-hectare plot of land along the road for the construction of a school and a town square. CREA

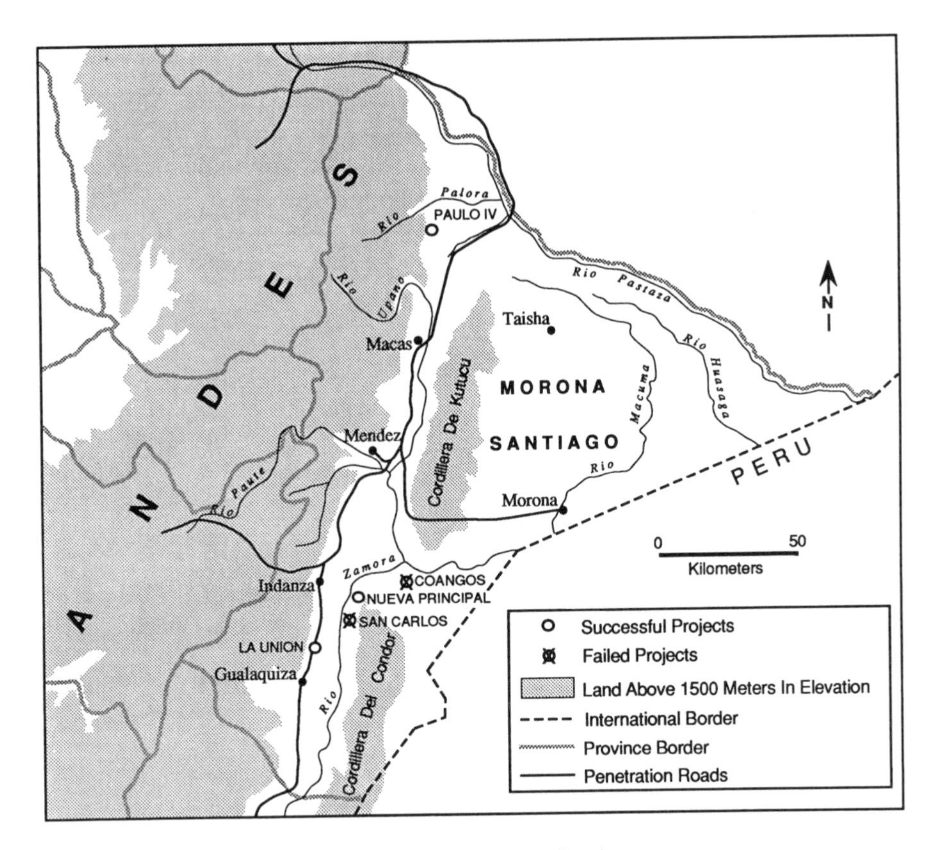

Figure 4.2. Colonization Projects in Morona Santiago

officials lent the brothers some road construction equipment to level the square. When the national government agreed to construct a school in the new village, many of the families who had purchased farms from the brothers bought lots near the school where they built houses for use while their children attended school.[14] The growth of an urban center with rudimentary services anchored the agricultural population in the area and gave them reasons to clear more land.

San Carlos de Zamora—A Failure. In 1968 a cooperative of Azuay peasants organized by a Peace Corps volunteer laid claim to a tract of mountainous terrain on the eastern bank of the Zamora River, two days' walk southeast of Indanza (see figure 4.2). The construction of a road from Azuay to Indamza had increased the value of the land and the numbers of people competing for it. The competition was fierce. A group of shopkeepers from Azuay had laid claim to a large tract of land next to the cooperative's claim, and the Ulloa brothers from Limon disputed both groups' claims to the land. To enforce their claim, the brothers attacked the shopkeepers' workers with machetes early one morning, and for the next three days the brothers shot at anyone who tried to enter the disputed area.

Twenty years later only four families lived in the community, and the land remained forested except for a few patches of cleared land along the river. Most colonists had given up, either abandoning their claim or selling it to one of the few remaining families. The mountainous terrain made it difficult to pasture cattle in many areas, which in turn reduced the size of the colonists' herds and the stream of income they could expect to earn from the land. The community's continuing isolation also worked against its success (see figures 4.3, 4.4, and 4.5). The absence of other settlements near San Carlos meant few local opportunities for wage labor, so colonists had to leave home to find work. The isolation also raised the cost of supplies for the settlers. The colonists had to pay the freight charges from Indanza to San Carlos in addition to the cost of goods. One mule carrying two hundred pounds of goods would cost a colonist three days' wages. No one planned to build a feeder road into the San Carlos region, so the colonists could not expect any relief from their burdens in the near future.[15] These constraints gradually convinced most of the colonists to abandon their claims.

Figure 4.3. On the Trail to San Carlos (photo by M. Gildesgame)

Figure 4.4. Mules Hauling Cargo near Indanza (photo by M. Gildesgame)

Nueva Principal—A Success. Because Nueva Principal is located near San Carlos, on the east bank of the Zamora River, its success offers an interesting point of contrast to San Carlos' failure. The colonists in Nueva Principal, originally from the same highland parish, settled first in San Clemente, a village on the west bank of the Zamora. Several years later, a group of them moved across the river and founded Nueva Principal. For several years relatives in San Clemente supported the colonists with goods and labor. The colonists' familiarity with the local environment enabled them to exploit supplementary sources of income more effectively than the newcomers in San Carlos. Initially, the colonists in San Carlos did not pan for gold in the Zamora; the colonists in Nueva Principal integrated gold mining into their daily round of activities. Whenever the women descended the steep walls of the Zamora's canyon to wash clothes, they would take their skillets with them and pan for gold after washing their clothes. With supplementary sources of income and support from extended families, the colonists in Nueva Principal made some economic progress. By the

Figure 4.5. Street Scene in San Carlos (Photo by M. Gildesgame)

mid-1980s enough families had settled in Nueva Principal to fill an elementary school with more than fifty pupils.[16]

Coangos—A Failure. The director of colonization programs at CREA started the Coangos project in the mid-1970s. The new settlement lay just north of the disputed international border with Peru, and the rationale for the project stressed the need to secure sovereignty over the area by settling it with Ecuadorian citizens.[17] The government constructed an airstrip and a network of trails in the area, but the rough terrain and its isolation discouraged the first settlers (see figure 4.2). CREA did not want to construct a branch road into the valley because it would have required building a large bridge and eighty kilometers of road over rough terrain. Without a connection to the national road network, all agricultural products would have to leave Coangos by air, and the cost of air transport from Coangos made it impossible to sell almost any product, including cattle, at a profit. When this grim calculus became apparent to most settlers, they left, and the project failed.

Paulo VI / 24 de Mayo—A Success. When colonists entered this region, just south of the Palora River, they received support from numerous institutions. The Salesians identified the unclaimed lands. CREA funded small construction projects in the new settlements, and the agency's directors promised to build a road into the settlements sometime in the future. Peace Corps volunteers measured the colonists' lands, and USAID, working through the Catholic Relief Service. provided free flour and powdered milk for the colonists' families for several years. Initially, each colonist received 30 hectares of land, but many of them later acquired an additional 60 hectares through an association claiming an adjoining tract of land. Credit on concessionary terms from the Banco de Fomento made it possible for the colonists to purchase small numbers of cattle, and the sale of cattle became the chief source of income for most colonists.

The remote location of the two settlements caused problems for the colonists. To reduce their isolation, the colonists began the grueling job of building an airstrip immediately after their arrival. For three months they felled trees and rooted out stumps on the tract of land which became their airstrip (see figure 4.6). For ten years the light planes using the airstrip provided an expensive, but direct link to the outside world. Road access to the region gradually improved as the construction crews, delayed by fiscal crises, pushed the head of the road north from Macas. When the road finally arrived in 24 de Mayo in 1989, nineteen years after the initial settlement, each village had more than five hundred people, and extensive tracts of cleared land surrounded both communities.[19]

Of the five attempts at agricultural expansion reviewed here, the outcomes at La Union and Coangos are perhaps least surprising. When agencies cannot provide important infrastructure, as at Coangos, the projects fail. When agencies like CREA and families like the Lopes provide the infrastructure, projects succeed. CREA followed this formula for success during the 1970s, and its efforts initiated a wave of land occupation and deforestation throughout Morona Santiago. The outcomes at Paulo VI and Nueva Principal seem more surprising. The colonists founded these communities in isolated places, and no one promised the quick construction of a penetration road. In spite of these disadvantages the two communities survived. The strength of the colonists' coalitions in the two places provides the most persuasive explanation for the communities' survival. In Paulo VI a coalition of public and private agencies supported the colonists. In Nueva Principal strong kin coalitions supported the first colonists. In both places the creation of a complex social organization offset to some degree the initial disadvantages of location and made it possible for the colonists to clear large areas of rain forest. The projects also succeeded because the colonists did not have to dislodge angry Shuar from the lands targeted for settlement.

THE SHUAR: ACCOMMODATION, RESISTANCE, AND CHANGE

Although the Shuar have had a well-deserved reputation for hostility towards outsiders, they have at various times developed working relationships with colonists. The Upano valley Shuar established peaceful relations with the mestizos in Macas during the nineteenth century because the mestizos could provide the Shuar with machetes and shot-

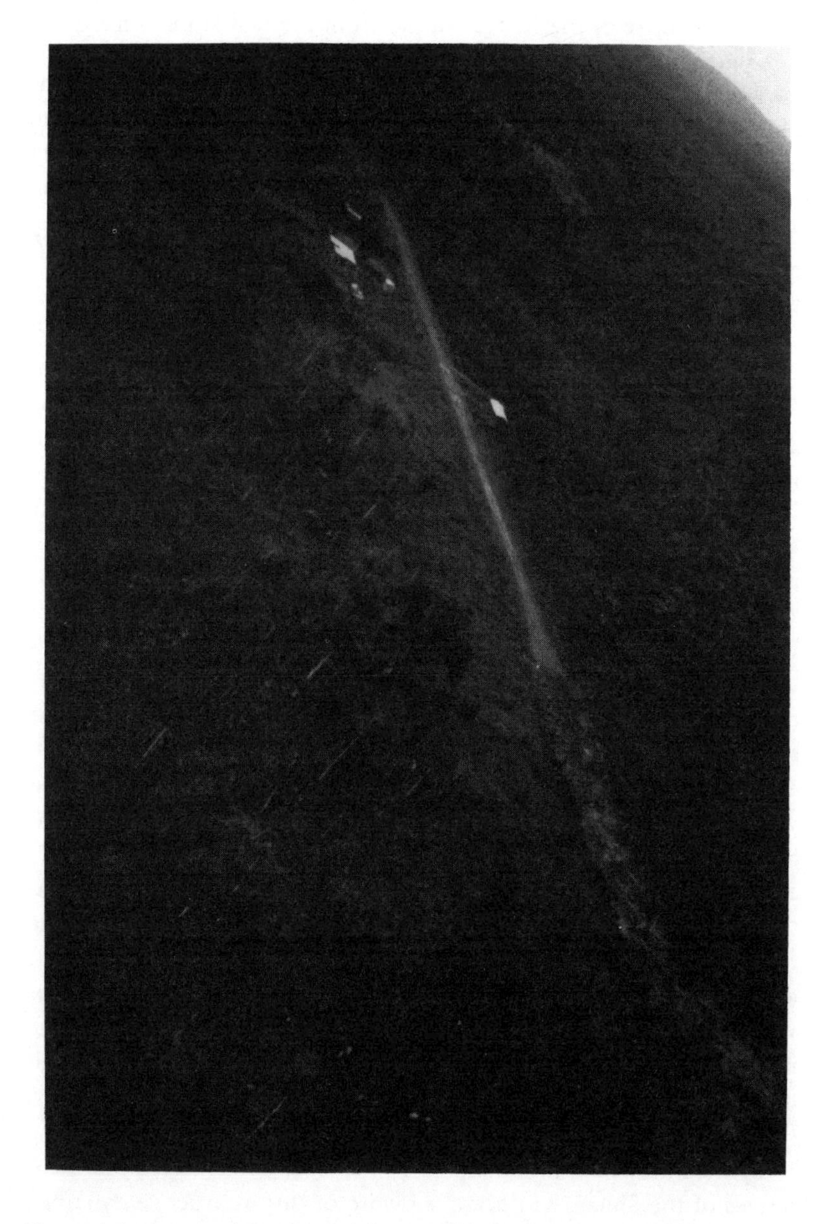

Figure 4.6. Approaching the Airstrip at Pablo Sexto (photo by M. Gildesgame)

guns that could then be traded with other Shuar living farther to the east. Contact with the first colonists from the highlands offered the same advantages to Shuar. As the colonists became more numerous during the 1930s and 1940s, their demands for land began to disturb the Shuar. Because the Shuar and the colonists made such different uses of the land, the outcomes of their struggles over land affected rates of deforestation. The colonists converted as much forest to pasture as possible; only the steepest slopes remained forested, primarily because cattle on hillsides could hang themselves on the ropes to which they were tethered. In contrast the Shuar practiced shifting cultivation which left the basic structure of the forest intact.

The dilemma facing two Shuar households in the valley of the Rio Indanza in 1952 typified the Shuars' situation during this period. The government decided to construct an improved mule trail between Indanza and Gualaquiza which would cross the valley of the Indanza River near the Shuar houses. Only two Shuar families lived in the affected area, a common situation given the group's dispersed settlement pattern. Prompted by the government's investment in infrastructure, the local Salesian priest, Luis Carollo, and the mayor of Indanza carved up a portion of the valley into small farms for seventeen or eighteen households of recently arrived migrants from the Sierra.[20] In the face of such numbers overt opposition by the two Shuar families did not appear feasible. Economic calculations also influenced the Shuar in such situations. By selling the land, they obtained valued material goods for consumption or for trade with the interior Shuar (Bottasso 1982:112). They may have calculated that they could easily obtain lands in adjacent areas. Large declines in Shuar population, caused by homicides and epidemics during the first half of the twentieth century (Harner 1973:33–35; Hendricks 1986), had left ample lands to the north and east to which the Shuar could move after vacating an area invaded by colonists. Between 1920 and 1960 large numbers of Shuar in the Limon–Indanza region sold their lands and moved farther into the rain forest in the face of colonist incursions (Federacion Shuar 1976:93; Harner 1973:36).

By the mid-1930s the Salesians, as the legally constituted representatives of the Shuar, had become aware of this destructive dynamic, and they took measures to insure the Shuar's continued access to rain forests. They persuaded the national government to declare large areas of western Morona Santiago as reserves for the Shuar. In some places the creation of reserves occurred in a confused, rumor-filled atmosphere that heightened tensions between colonists and the Shuar. As one Salesian priest recalled,

> problems could have been avoided if Padre Juan Vigna, then provicariate of the Mission, in requesting the Reserves, . . . could have surveyed the particular places and populations involved in each reserve. He worked alongside Dr. Luis Octavio Diaz, then director of the Oriente, and it appears that everything was done in haste. One had to take advantage of opportunities, so everything was done in a simple and precipitous way. For example, the Reserve of Zarambiz had such imprecise boundaries that it appeared to include the entire [mestizo] town of Limon. Thank God that Padre Juan Schmid [of the mission in Limon] was able to explain the situation, quell the emotions, and then proceed to solve the problem.[21]

In other places, such as Huambi, the controversies did not occur until decades after the creation of the reserves when growing colonist populations, unable to find land near their homes, began to invade the forested reserves.[22] Successive colonist invasions of the Shuar reserve south of Sucua gradually reduced it to one-third of its original size (Hendricks 1986:37).

In the 1950s in an attempt to secure the Shuar's hold on the land, a priest helped the Shuar near Limon obtain individual titles to land. This plan went awry when many of the new landowners sold their lands to colonists (Bottasso 1982:116). Under these circumstances the Salesians began to promote the idea of global titles to land. The Shuar in an area would form a *centro*, an organization of villagers, and it would receive title to a large tract of land around the village. Each household in the village would receive a tract of land in the *centro*. Household heads could sell their land to other members of the *centro*, and they could pass it on to their sons and daughters, so individuals considered themselves to be "owners" of their tract of land. They could not sell their land to outsiders.

In the 1960s the Shuar usually formed *centros* when colonists threatened to take over their land. Colonists would begin clearing land in an area containing unorganized Shuar families, and the Shuar would respond by forming a *centro* and filing an application for a global title to the surrounding land.[23] In some instances the Shuar did not face an immediate threat from colonists, but they wanted to establish pastures and gardens, so they asked the priest to divide up their land and establish a center.[24] In many instances the formation of a *centro* precipitated a change in the settlement pattern. The Shuar in a *centro* abandoned their isolated homes in the forest and built new homes in a village center.

In 1964 at the urging of the Salesians, the Shuar formed a federation of *centros*. The Federacion de Centros Shuar enabled the indigenous peoples to speak with one voice and increased their power in struggles with colonists. In the late 1960s the federation began an aggressive program of acquiring global titles to land. At first federation officials focused on securing a land base in disputed areas west of the Kutucu; later the federation promoted the formation of *centros* in undisputed, unorganized areas east of the Kutucu.

To obtain a title, the Shuar had to form a village in the claimed area. In most instances the priests expedited the organizational process by visiting all of the Shuar households in an area. During these visits the priests urged the Shuar to form a *centro* and begin cattle ranching on individual tracts of land around the village. If a proposed center had relatively few inhabitants, the priests would encourage Shuar from other areas to migrate to these *centros*.

Rapid population growth after 1950 among the Shuar living around the mission settlements in the Upano valley had created land shortages, leaving many of the younger Shuar without land.[25] In some instances these Shuar decided on their own to leave the Upano valley; they settled among relatives or acquaintances in sparsely populated areas, a process of spontaneous colonization.[26] In other instances the priests provided incentives. They would promise young, Upano valley Shuar a piece of land (60–70 hectares) in a new *centro* to the east of the Kutucu range if the Shuar would move there.[27] In this fashion the federation created centers and secured land for the Shuar even in places, all to the east of the Kutucu range, where the local Shuar population might not have been large enough to form a village. As at least some Shuar leaders recognized, the Salesians and the Shuar had created their own distinctive colonization program.[28]

While colonist invasions provided the impetus for change in Shuar society west of the Kutucu between 1920 and 1960, the federation became the chief stimulus for change east of the Kutucu during the 1970s (Hendricks 1986:209). The speed with which the Shuar secured claims over large tracts of land east of the Kutucu alarmed officials in CREA's colonization program. In the words of one official, the Shuar "with technical assistance in surveying from foreignvolunteers, German as well as Italian, are taking large blocks of land for families which in the best of cases do not number more than 30, reserving in this manner enough physical space for the fourth generation to come!"[29]

To prevent the Shuar from establishing more legal claims to land east of the Kutucu mountains, CREA's directors persuaded the military junta to issue decree 3134-A in 1977. The decree established a national reserve for CREA's colonists east of the Kutucu mountains. While CREA's colonization programs were technically open to the Shuar, CREA had never assisted the Shuar in forming settlements, so decree 3134-A appeared, in effect, to reserve an extensive area of Morona Santiago for settlement by mestizos.[30] The Shuar Federation lobbied vigorously against the decree, pointing out that the reserved area already contained some eighty Shuar settlements and should therefore be considered Shuar land.[31]

During the same period CREA officials may have put pressure on bureaucrats at the agrarian reform institute, IERAC, to delay the processing of Shuar land claims, thereby slowing down the expansion of Shuar controlled territories. Statistics on the legalization of land titles support this claim. Between 1972 and 1975 IERAC legalized Shuar claims to approximately 95,000 hectares of land. In the following four-year period, 1976–1980, IERAC legalized only 33,000 hectares of Shuar claims, and the agency's backlog of unresolved Shuar land claims grew rapidly.[32] When control over the national government shifted to left of center parties in 1988, IERAC's titling policies changed, and the agency approved most of the Shuar's longstanding claims to land.[33]

Figure 4.7 and table 4.2 outline the pattern of Shuar landholding which emerged from the sixty-year struggle with the colonists. Table 4.2 reports the proportion of arable land held by the Shuar in five subregions of Morona Santiago.[34] Two patterns stand out. First, the Shuar have retained a larger proportion of the land in the more recently settled regions. This pattern suggests that the Shuar, through the

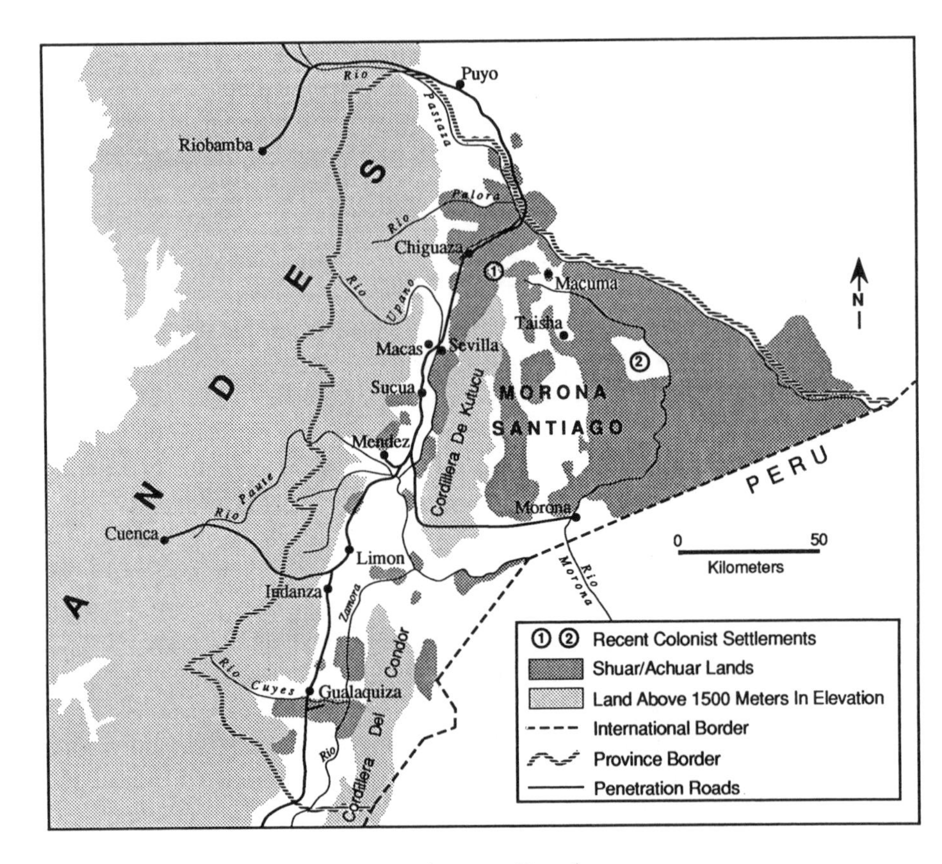

Figure 4.7. Shuar Lands in Southeastern Ecuador

Table 4.2. Shuar Landholdings in Morona Santiago

Region	Period of Settlement	Proportion of Arable Land Held by Shuar
East of Kutucu	1965-present	83%
Upano—Palora plain	1955-1985	53%
Upano Valley	1920-1970	38%
Limon–Indanza	1910-1970	8%
Gualaquiza	1910-1970	34%

SOURCE: Mapa, Ubicacion de Centros, Federacion de Centros Shuar, 1985.

federation, became more effective in establishing their claims to land. Second, the variations in the extent of Shuar landholdings in the three oldest colonization zones indicates how natural features can influence settlement processes. In both the Upano valley and the Gualaquiza area impassable or intermittently impassable rivers bisect the regions; the Upano River cuts through the former region; the Cuyes-Bomboiza and Zamora rivers run through the latter region. The difficult river crossings prevented rapid colonization of lands to the east of these rivers and in effect converted these lands into regions of refuge for the Shuar (Aguirre Beltran 1979).

In the Limon–Indanza region, almost all of the lowlands lay west of the Zamora River, so the Shuar had no refuge, and colonists obtained a much larger proportion of the lowlands than they did in other areas of Morona Santiago. In several instances in the Gualaquiza and Upano valley regions the Shuar deliberately withdrew to one side of a river.[35] The difficulties of getting cattle across a raging torrent like the Upano discouraged the colonists from trying to establish farms on the far side of a river. While colonists could change a line of stakes in a forest, they could not change the course of a river, so rivers provided the Shuar with an unambiguous, easy to police boundary with the colonists.

In at least one instance the Salesians added political obstacles to the natural barriers created by the rivers. They opposed construction of a bridge over the Cuyes River out of fears that colonists would cross the river and encroach on Shuar lands (Eckstrom 1979:111). In this manner the Shuar retained control over these lands until the 1970s when they obtained legal titles to the land.

Unlike the western regions, the area east of the Kutucu still contains considerable amounts of unclaimed land located between Shuar villages (see figure 4.7). Rugged topography and the pattern of growth in Shuar villages accounts for the unclaimed land. Between 1975 and 1985 the number of Shuar *centros* increased from 139 to 243.[36] Almost all of the new centers formed when demographic pressures or political strife caused a split within an old center. One of the two groups then occupied an adjacent area of unclaimed land. In some instances Shuar would not occupy the unclaimed land between their villages because the heads of households from different villages did not trust each other enough to live together in a new *centro* on the interstitial land.[37] In two instances during the 1970s colonists took advantage of these cleavages between groups of Shuar to claim and clear extensive tracts of interstitial land. Colonists from Sinai claimed land and established a community (Ebenecer) in the area between the Chiguaza *centros* organized by Catholic missionaries and the Macuma *centros* organized by Protestant missionaries (the area marked ① in figure 4.7). Colonists from Taisha claimed and cleared an extensive area between the Panki and Macuma rivers ② in figure 4.7), which had previously served as a buffer zone in the hostilities between the Shuar west of the Panki and the Achuar east of the Macuma. The light areas above the cross in figure 4.9 indicate the Panki-Macuma clearings.

Although the Shuar obtained titles to large tracts of land during the early 1970s, they did not relax their efforts to secure their land base. Violent conflicts with colonists over forested Shuar lands in the Upano valley, coupled with the new emphasis on intense land use in Ecuador's agrarian reform law, convinced the federation's leadership that titles to land would not insure control over it. The 1973 agrarian reform law gave legal force to the colonists' norm that "he who works the land, owns it," so the Shuar felt that they had to intensify their use of land in areas bordering colonist settlements if they wanted to retain control of it.[38] To this end, the Shuar decided to work the land just as the colonists did, by becoming cattle ranchers. Credit to acquire cattle came from two sources. In the early 1960s the Salesians formed a "bank of the poor" through which they lent cattle to the Shuar and allowed them to keep every other offspring.[39] In the early 1970s the federation, using funds donated by European development agencies, began making loans to Shuar *centros* for the development of their cattle herds.

To receive a loan, each *centro* had to form a cattle development group to administer the loan. In pursuit of economic gain, these groups accelerated local deforestation. This dynamic was especially evident in the *centros* east of the Kutucu. The groups built small jungle airstrips to evacuate sick villagers and to transport slaughtered cattle to markets in the western Oriente. The access to markets provided by small airplanes encouraged the residents of villages with airstrips to develop their cattle herds. While only 47 percent of the *centros* east of the Kutucu had airstrips, 86 percent (six of seven) of the eastern *centros* that obtained cattle development loans had airstrips.[40] To feed their growing herds, the Shuar in these *centros* cleared land and planted pasture. By the mid-1980s the interior villages with airstrips had more cleared land, 6.8 percent, than the villages without airstrips, 1.9 percent ($p=.01$).[41] By building infrastructure and acquiring capital, the cattle development groups, like other growth coalitions, accelerated rates of deforestation in the interior villages.

By forming *centros* and committing themselves to cattle ranching, the western Shuar abandoned their dispersed settlement pattern and their forest-based subsistence cycle. They began to resemble their adversaries and, as Simmel (1955) would have predicted, the growing similarity in the two ethnic groups' goals intensified the conflict between them. With these changes the significance of the struggle between the western Shuar and the mestizos for the preservation of rain forests began to change. Indigenous control of the land no longer meant that the land would remain forested.

The federation's political activities reflect the change in the western Shuar's interests. In the 1980s, at least in the western portions of Morona Santiago, the federation began to advocate economic development. They pressured the state to build a hydroelectric project, maintain airfields, and construct bridges in selected Shuar communities.[42] In one western community the Shuar sold some land to a tea company after calculating that the exchange value of their remaining lands would increase when the company built an access road into the newly acquired lands (Eckstrom 1979:138). At the same time the Achuar and the interior Shuar east of the Kutucu range, still living for the most part in forests, opposed the construction of a new road into their region because, by bringing colonists and development projects into the region, it would destroy the forests that sustain their way of life.[43] Caught in between the acculturated Shuar in the west, who wanted some development, and the unacculturated Shuar in the east, who wanted to preserve their forest based subsistence cycle, federation officials argued that roads had their "advantages and disadvantages."[44] These regional differences in attitudes about the forest suggest that ethnic divisions provide only a partial guide to the geography of deforestation in Morona Santiago.

THE GEOGRAPHY OF DEFORESTATION

The different attitudes toward development among the eastern and western Shuar reflect differences in land use within Morona Santiago during the 1980s. Figure 4.8 outlines the historical geography of deforestation in Morona Santiago.[45] Table 4.3 presents analyses of a 1983 satellite image that demonstrates the variation in land cover across the province. In each of the three areas, we classified the land cover in representative subareas. A Shuar center would make up one subarea and a colonist zone of approximately equal extent made up another subarea. We analyzed seven subareas in the Upano valley, twenty-one subareas in the Upano—Palora region, and ten subareas east of the Kutucu range.[46] We excluded town centers and land with rugged topography from the analysis; these procedures make the deforestation data comparable across regions, but they also elevate the overall estimates of deforestation given that mountainous terrain, almost always forested, has been excluded from the analysis.[47]

In the Upano valley, the oldest settlement area analyzed, Shuar and colonist communities do not differ significantly in the extent of deforestation. The uniformity in land use reflects both political and demographic pressures. Between 1930 and 1970 large blocks of Shuar-controlled rain forest became the focus of repeated controversies, the last of which occurred in 1970 when disgruntled colonists burned the Salesians' mission house in Sucua. In many instances the colonists acquired a portion of the disputed lands and cut down the forest. Trends in census data across both ethnic groups reinforces the argument that, if political pressure does not produce deforestation, the pressures of a growing population will accomplish the same end. In the old colonization zones of Mendez and Limon which were settled

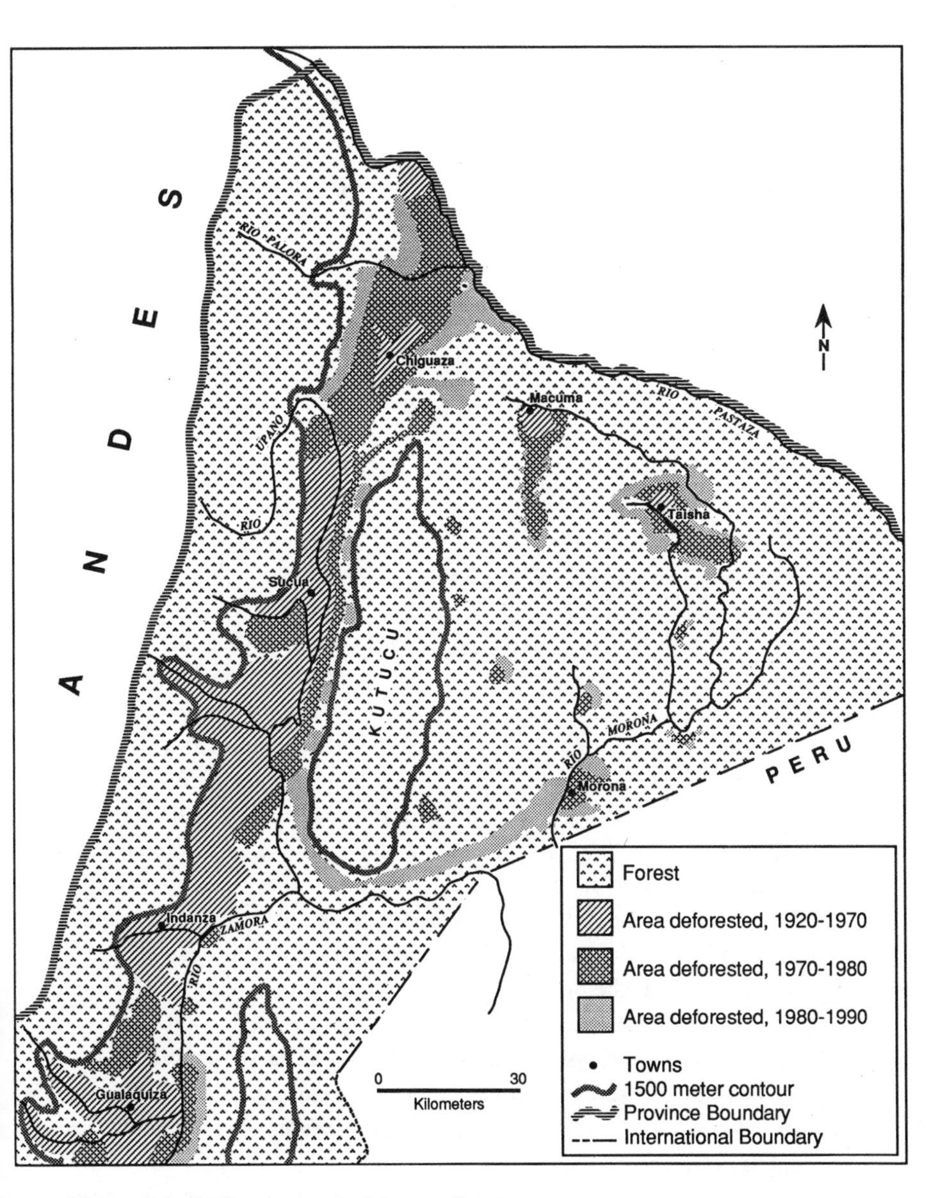

Figure 4.8. Deforestation in Morona Santiago

Table 4.3. Deforestation in Morona Santiago, 1983

	Land Cover – % Forested		
Region	Mean	Shuar	Colonist
East of Kutucu	91.9%	94.9%	70.0%
Upano—Palora Plain	52.8%	70.5%	33.4%
The Upano Valley	6.3%	7.1%	5.0%

SOURCES: Multispectral Scanner Image, 1983; Mapa: Ubicacion de Centros, Federacion de Centros Shuar, 1985.

nearly seventy years ago, the modal property now contains 15 to 20 hectares in both colonist and Shuar communities. Fields and pastures in these areas have climbed the hillsides as settlers and the Shuar have struggled to bring more land under cultivation. Farther north, in the more recently settled colonization zones of Macas and Sucua the modal property is larger, 40 to 50 hectares, and people have not brought steep slopes under cultivation (Ministerio de Agricultura — Pronareg 1980b:11). This pattern suggests that population pressure and the related subdivision of properties forces both colonists and Shuar to intensify their use of land and eliminate the last islands of forest on their land.

Rapid deforestation did not begin in the Upano—Palora region until the late 1960s when the impending construction of the Macas–Puyo road spurred settlement in the region. By the mid-1980s colonists and Shuar had claimed all of the arable land in the region, but much of the land far from roads remained forested. As the data in table 4.3 make clear, ethnicity plays an important role in land clearing in this region, with the colonists cutting down the forest much more rapidly than the Shuar.

The most intense competition for land in the early 1990s occurs east of the Kutucu range. Both colonists and Shuar from the Upano valley have begun settling around Taisha where the Salesians have a large mission and the military maintains a garrison at an airport. The colonists and the Shuar in the immediate vicinity of Taisha have cleared about 30 percent of their land. The light area above the center of figure 4.9 indicates the clearings around Taisha. Even the *centros* east and south of Taisha, with approximately 5 percent of their land cleared, show some signs of deforestation.[48] As noted earlier, the vil-

Figure 4.9. Land Cover East of the Kutucu, 1983 (source: LANDSAT)

lages with airstrips have cleared more land than comparable villages without airstrips. The places without airstrips often contained unrelated families, grouped together in artificial centers (Hendricks 1986). The families chose not to work together, and their lands remained forested. These *centros* have had difficulty obtaining titles, and their land has become the target of colonization schemes by both Shuar and colonists. With the completion of the Mendez–Morona road into the southern portions of the region in 1988, groups of colonists with plans to occupy these areas multiplied, and rumors filled the air about a multinational company, attracted by the fertile lands, which wanted to establish an African palm plantation in the region.

While the competition between the Shuar and the colonists for control over land explains much about local variations and short-term fluctuations in deforestation rates, the long-term eastward expansion of settlement depends in large part on the variable activities of growth coalitions and lead institutions. The comparison of colonization in the Cuchibamba and Cuyes valleys between 1920 and 1970 establishes the importance of a lead institution, the missions, in starting settlements and determining the rate at which they grow. Comparisons of more recent colonization projects suggests the importance of growth coalitions and an agency's road building to the success of efforts to claim and clear land. Some of these patterns are less obvious than others. The crucial role of growth coalitions, their wide variety and fleeting nature, only becomes apparent in detailed histories of the struggles to settle particular places. The following chapter tells one of these stories. It illustrates the difficulties of trailblazing and the value of growth coalitions in overcoming obstacles to settlement.

5

Trailblazing in a Large Forest: The Upano—Palora Plain in the 1960s

RIVER CROSSINGS AND THE FIRST SETTLERS

For hundreds of years a forest stretched away to the north and east across the Upano River from the town of Macas. Farmers on the river's west bank never extended their farms to the east bank because they did not want to cross the river. Like all rivers in the southern Oriente, the Upano cuts deeply into the surrounding plain. High cliffs run the length of the river on both sides; the bed of the river, composed of volcanic rock and sand, extends approximately one kilometer from cliff to cliff. The river runs down this sandy spit in multiple channels with many rapids (see figure 5.1). The runoff from major storms carves new channels and abandons old channels, so the channels change frequently (see figure 5.2). A backwater can turn into a raging torrent overnight.

To cross the river prior to 1970, travelers would wade knee and waist deep across minor channels and climb into a dugout canoe to cross the major channel. Crossings were hazardous undertakings; the river at Macas regularly claimed one or two lives a year during the 1960s; countless cattle and horses also lost their lives when their owners forced them to ford the river. These hazards made the Upano a formidable barrier to settlement. Except for widely separated Shuar houses no one lived east of the river until the mid-twentieth century.

The Shuar and the Salesians established the first permanent settlements on the east bank of the Upano during the 1940s. During the 1920s and 1930s Shuar graduates of the Salesian schools in Macas began to establish homes and gardens in Sevilla Don Bosco (Sevilla), directly across the river from Macas. In 1943 the Salesians transferred their boarding school for Shuar children from Macas to Sevilla. In 1954 the Salesians founded a mission at Chiguaza, a day and half's walk north of Sevilla (see figure 4.1). Large numbers of Shuar living around Chiguaza had sent their children, beginning at age eight, to the Salesian boarding school in Sevilla. The long separations disturbed both parents and children, so the parents began to lobby for a mission school closer to their homes. The Shuar in Sevilla, burdened by a recent increase in population, complained about poor hunting and a lack of land for gardens close to the mission. The Salesians responded to these complaints by opening a new mission in Chiguaza and persuading a number of Shuar from Sevilla to form a community on uninhabited lands near the new mission.

Several years later two wealthy farmers from Macas established the first pastures for cattle east of the Upano. The farmers' initiative stemmed from problems they had in getting their cattle to markets. In the late 1950s ranchers in the Upano valley drove their cattle south to Mendez and up the Pan-Mendez trail to the Sierra, a punishing trip requiring more than fifteen days of travel. Cattle merchants from Puyo, 120 kilometers to the north of Macas, suggested an alternate way to market cattle. If the ranchers from Macas could create pastures on the east side of the Upano, their cattle would not have to cross the Upano to get to the market in Puyo. When the cattle were ready for sale, they could be driven north to Puyo, a trip that would require only two major river crossings, one of which, at the Palora, was not as hazardous a crossing as the Upano. Following the merchants' suggestion,

Figure 5.1. The Upano River North of Macas (photo by M. Gildesgame)

the two ranchers began clearing land across the river from Macas. Other ranching families followed suit, claiming and clearing tracts of land along the trail north to the Palora River. The colonists established their claims in increments. They would befriend a nearby Shuar and purchase a small piece of land from him. Over the years through a series of purchases, the colonists' farms grew larger.[1]

The colonists sent their youngest sons to work on the new farms, or they recruited a tenant farmer to live on the farm, watch the cattle, and clear land. For his work the tenant received every other offspring from the herd (an *a medias* agreement). In its northern sections the trail ran through lands held by the Shuar; they too cleared small amounts of land and rented the pasture to peasants driving herds north. By the mid-1960s a corridor of pasture, visible in remote sensing images twenty years later, ran the length of the trail. Unbroken expanses of forest covered the plain to the east and west of the corridor.

In the mid-1960s the Upano valley still did not have a road connecting it with population centers in the Andes, but CREA had obtained a loan to build a penetration road from Azuay into the

Upano valley, and they talked about securing a second loan to continue the construction of the road north to Puyo (see figure 4.1). CREA went so far as to carry out preliminary studies for the second road. The construction of a bridge over the Upano posed special problems for the engineers. The shifting course of the Upano argued for building a bridge from cliff to cliff, but the long distance between the cliffs in most places would have made this type of structure extremely costly. Initial engineering studies indicated that the most feasible point of crossing would be a place called "The Narrows," twenty-five kilometers north of Macas, where the river emerges from the mountains, and the distance from cliff to cliff is only one hundred meters (see figure 5.2). The choice of this route by CREA became common knowledge in Cuenca during the mid-1960s, and two groups of investors from Cuenca decided to establish large cattle ranches in the path of the future road, east of the narrows, on the Upano—Palora plain. In 1967 workers from the two groups, headed by C. Duran and J. Monsalve respectively, established camps, cut boundary lines, and began to clear the rain forest on their claims.

The Duran group made the larger investment. Carlos Duran established a residence on the east bank of the Upano, hired twenty workers to clear land, and strung a cable over the largest branch of the Upano to insure access to his ranch (see figure 5.3). Duran claimed a large area of rain forest, but he never knew its exact extent. When he first cut the boundaries, he estimated that he had 3000 hectares; later, after a fracas with the agrarian reform agency, he lowered his estimate to 1500 hectares. Several farmers from Macas established competing claims to land within the boundaries cut by Duran. During the first several years Duran eliminated his competitors by working all of the land around their clearings, thereby eliminating their claims to the surrounding forest.[2] With his competitors' clearings stripped of much of their value, Duran would offer them small sums of money if they would renounce their claims to his land. The rivals always took the money and abandoned their claims.

Both Duran and Monsalve maintained amicable relations with nearby Shuar. Duran, Monsalve, and the colonists occupied land above 1100 meters in elevation, and the Shuar planted their gardens below 1100 meters, so the two groups never came into direct conflict. The large landowners and the colonists ruined Shuar hunting grounds

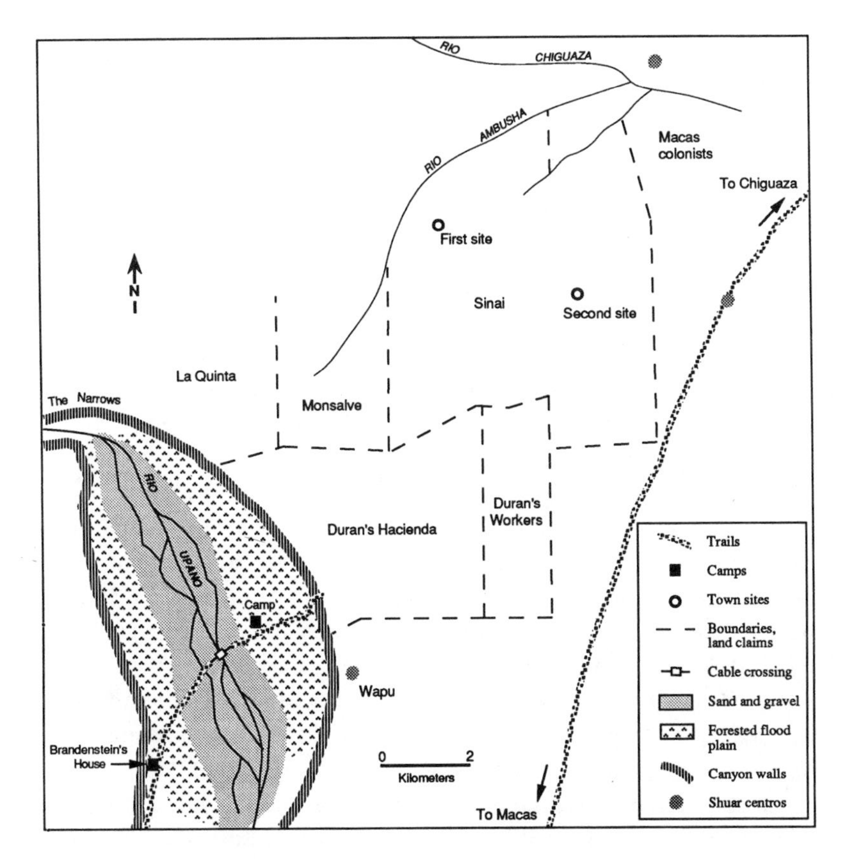

Figure 5.2. Land Claims, Southern Upano ---Palora Plain

by clearing these lands, but they did not occupy lands that the Shuar claimed for their *centros*, so relations between the two groups remained peaceful.

TRAILBLAZING IN THE RAIN FOREST

The first group of CREA-assisted colonists arrived on the west bank of the Upano in 1969, two years after Duran entered the region. Their involvement with the colonization project began six months earlier when promoters from CREA and the Peace Corps began organizing groups of potential colonists in their highland communities. As inducements to settle in the Oriente, CREA offered the colonists technical assistance and foodstuffs during their first year in the rain forest. Like spontaneous colonists, CREA's colonists had to locate and develop the lands themselves.

Although CREA's promoters tried to recruit colonists from Azuay's poorest communities, most of the colonists came from more prosperous places.[3] The would-be colonists were poor, but not the poorest people from their communities. Approximately 50 percent had no land; 45 percent had less than one hectare, and 5 percent had more than a hectare.[4] Once formed, the group began to look for vacant lands in the Oriente. After an arduous trip to explore an uninhabited valley three days' walk east of Limon, several representatives of the group made a trip to look for lands north of Macas. The advance party reached the cable crossing on the Upano, but CREA officials had not consulted with Duran, so the colonists could not use his cable to cross the river and look for unclaimed lands (see figure 5.2). From conversations with Duran's workers the colonists learned that most of the land across the river did not have owners and, from afar, they could see that the land was flat, so they decided to settle on it. With only these general guidelines to follow, the main party of colonists reached the banks of the Upano several months later.

To get to the Upano, the colonists traveled for four days. From their homes in the Sierra they rode in the back of dump trucks to the end of the penetration road in Limon. The next day they walked to Mendez, forty kilometers to the north. The following morning they walked north another fifteen kilometers to the head of a road where

Figure 5.3. Duran's Cable Over the Upano River (photo by M. Gildesgame)

they boarded trucks that carried them to the end of the road north of Macas. From the end of the road they walked another six hours the next day into the house of Helmut Brandenstein, across the Upano River from Duran's hacienda (see figure 5.2). Brandenstein, a self-described former Luftwaffe pilot, had come to this remote place in the late 1940s and started a farm. In May 1969 his house above the Upano became a base for approximately forty colonists in search of a sizable chunk of unclaimed land on the far side of the river.

The setting had a menacing air about it. To the east ran a wild river, the Upano. To the west and northwest the black, volcanic wall of the Andean escarpment, capped by Sangay's volcanic cone with its constant trail of smoke, rose 10 to 14,000 feet above them. While the unusual surroundings fed the colonists' apprehensions about the value of their undertaking, most of their concern focused on the continuing uncertainty about the location of the lands they would occupy. To resolve this question, the group formed a small exploration party to find land.

The first group of explorers crossed the river and traveled north along the cliffs bordering the Upano. After locating unclaimed lands north of Duran's property, they tried to find an easy route from the cable crossing up the cliffs to the unclaimed lands. When they could not find a way up the cliffs, they rejected the idea of settling on these lands and returned to Brandenstein's house. Their failure to find suitable lands discouraged their compatriots, some of whom abandoned the group and returned to the Sierra. The ties of kinship and friendship that played an important role in the colonists' recruitment also influenced their decisions to abandon the colonization effort. When several colonists left the group after the failure of the first exploration trip, their friends and relatives went with them.

A second exploration party left several days later to explore lands to the northeast of the Duran property. Duran led the explorers across his lands to a small creek that marked the end of his property. The exploration party crossed the creek and, with the aid of a compass, began exploring the unclaimed land by walking north, then east, then south, and finally west to close a square. The peasants traversed the land in silence, wondering if they would get lost in the unfamiliar expanse of rain forest. They noted the slope of the land, the availability of water, and the presence of *cana guadua*, a bamboo, which indicated fertile lands to the colonists. The explorers returned to Brandenstein's house convinced that they had found suitable lands for settlement.

While the explorers looked for land, the colonists at the house, anxious to do something other than consume supplies, began clearing land on the floodplain east of the river (see figure 5.2). Although the land had only a thin cap of topsoil on top of gravel from the river bed, the colonists liked its location. So did other people. Shortly after the colonists began to clear the floodplain, three residents from Macas showed up and lodged a counterclaim to these lands. The Macas residents did not back up their claim by clearing land, but they eventually acquired the land anyway. After clearing almost 20 hectares of forest, planting corn, and building a large camp, the colonists abandoned these lands for the new claim northeast of Duran's hacienda. By the time the corn crop matured six months later, the colonists had forgotten about it, so the corn rotted in its stalks.

For several weeks after the discovery of the unoccupied lands, a small group looked for a suitable site for a permanent settlement. Although the colonists had just acquired a prized possession, many of them showed little inclination to look it over. They lived in fear of losing their way and spending a night alone in the forest, crouched at the base of a big tree, worrying about snakes, and waiting for dawn.[5] To avoid getting lost, the colonists kept to recently established trails and learned about the new place in small increments. Their caution influenced the choice of a site for the town. Unwilling to explore extensively and reluctant to move farther to the north and east, away from the road and their source of supplies, the colonists did not mount an aggressive search for a town site. After several short explorations, they chose a site less than a kilometer from their first camp. Four months later they abandoned this site for a second site, about fifty meters lower in elevation and four kilometers farther east. The decision to move the town center came after further explorations revealed an extensive tract of unclaimed land, extending to lower elevations some four to five kilometers east of the first town site. Although the new site's lower elevation made it easier to obtain water and grow tropical crops in backyard gardens, many colonists found it difficult to abandon the first site because it meant abandoning months of work on houses and gardens.

The highlanders' unfamiliarity with rain forests undoubtedly contributed to their initial difficulties in settling the land, but the history of other communities suggests that colonists with substantial experience in rain forests make similar mistakes. Colonists moved the town site of Indanza twice before deciding on its current location.[6] Colonists moved the town center of La Quinta cooperative twice in an attempt to find a dry, centrally located site with easy access to water. Even the Shuar made mistakes in creating *centros* because they did not know an area well enough. In establishing the boundaries of the Centro Uunt Chiwias, to the northeast of Chiguaza, the Shuar chose to follow the cliffs overlooking the Chiguaza River rather than the river itself. They assumed that only a few meters separated the base of the cliffs from the course of the river. Later it turned out that the strip of forest between the river and the cliffs was more than five hundred meters wide in some places. The initial confusion caused conflict later on. Mestizos moved onto the land during the 1970s, claiming that it lay outside the boundaries of Uunt Chiwias, and a dispute over ownership began.

Given the colonists' almost chronic disorientation, the first paths between places followed circuitous routes. Eventually someone would realize the roundabout course of a trail and begin arguing for a new, more direct route. The repeated relocations of the trail between the cable crossing of the Upano and the new town illustrate this process of trial and error. At first the colonists used Duran's trails to get from the cable crossing to the village center. After several months in the new town, the colonists tried, with the aid of a compass, to cut a more direct path through the forest to the cable crossing. With the inevitable detours around pastures and Shuar gardens, the new trail became a winding path. A third trail, with CREA paying the colonists to do the work, produced the most direct route to the river crossing.

The relocations underline the trailblazers' ignorance about the land they had occupied. Although CREA paid for a reconnaissance flight over the colonization zone, the passengers on the flight were the project's financial backers and not the poor peasants walking around on the forest floor, trying to decide where a boundary runs or where to establish a camp. In forests with thick canopies overlooks are difficult to find, so the first settlers rarely get the "big picture." Geographical knowledge advances in increments in this setting. A colonist makes a map one week, only to remake it a week later with new knowledge of his surroundings. Everyone makes and remakes maps. Peasants sketch crude maps on dirty notebook paper to convince coworkers about the proper direction for a trail or the extent of land between two rivers.

When colonists cut boundaries through the forest, they learn about their land and sometimes it contains surprises, like a competing claim to the land. A faint footpath in the forest often provides the first sign of a competing claim. Shortly thereafter, the colonists encounter a boundary, a line of stakes stretching off through the vegetation in the shadowy light of the forest floor. The moment of discovery has an electric quality to it. The empty forest suddenly becomes inhabited, and the adrenalin flows. Crew members expect the rival claimant to step out from behind a tree and warn them to go no farther. Invariably he does not, but the discovery of the boundary precipitates a quick search to see if the competitor has backed up his claim by clearing land. If the colonists do not discover any clearings, they cross the boundary with their own boundary line, and a dispute begins.

In this fashion the colonists from Sinai crossed the boundaries of ranchers from Macas who had claimed tracts of land along the Upano—Palora trail (see figure 5.2). The families from Macas had cleared land along the trail but not away from it. They now found Sinai's colonists challenging them along their rear boundaries. Both parties enforced their claims by clearing extensive tracts of land in disputed areas. No one had title to lands in the Upano—Palora zone, but everyone acknowledged that "he who works the land owns it." A corollary to this axiom concerns wealth. He who can pay many workers to clear land will own much of it.

Dense forests, crisscrossing boundary lines, and subsistence crises slowed down the process of establishing secure claims to tracts of land. Working only with axes and machetes, colonists could not cut more than a kilometer of a boundary line during a day's work. The new settlement suffered from frequent subsistence crises, which slowed down the subdivision of the land. When supplies ran low, colonists would walk to Macas, the cable crossing, or Chiguaza, wherever food could be found, and load it on overworked pack animals or carry it on their backs into the new settlement.[7] In some instances the surveyors, all U.S. Peace Corps volunteers, would depart for the city in search of supplies or relief from life in a jungle lean-to. Their departure suspended the survey work. In innumerable instances the colonists would leave to work as wage laborers in order to replenish their families' cash reserves. Because the colonists had agreed that no one could receive land in absentia, a departure meant a delay in the receipt of land. With all these delays some colonists had wait six to eight months before they received a tract of land and could begin clearing the forest.

Along with delays the colonists had to endure the frustrations of false starts. In their first four months in the Upano—Palora rain forests they abandoned one month's work on the floodplain of the Upano and three month's work in the first town center. Only colonists with stout hearts and deep pockets could withstand these reverses. Almost all of these men had wives with children who lived on reduced incomes in precarious arrangements with relatives in the Sierra. Some dependents suffered so much in these arrangements that colonists felt compelled to bring them out to the new farm before they had harvested crops from their fields and gardens. Getting enough to eat became a constant preoccupation for these families during the next six months. The inadequate diets contributed to cases of anemia among the younger children.

The pressures of allocating desirable goods like land made it difficult to maintain peace within the cooperative at Sinai. The coalition's members came from different communities, and they met for the first time in the jungle. They did not trust one another, but they had to make important joint decisio about who would receive which tract of land, decisions with important implications for the individuals' futures. Under these conditions colonists from one Sierra community clashed repeatedly with colonists from another Sierra community over the allocation of land and food. These conflicts underscore the contingent nature of the coalitions; in the midst of strife they can fall apart quickly.

For approximately 60 percent of the colonists who left the Sierra, the privation, conflict, and uncertainty became too much to bear, and they abandoned the effort to colonize the rain forests. The first group of colonists to settle in Sinai entered the forest with twenty-five members; two months later it had only eleven members.[8] These hardships and the high rates of human retreat do not characterize land-use conversion around small islands of rain forest, but they do characterize processes of spontaneous and state-sponsored colonization in large blocks of rain forest.

So what sustained the Sinai colonists during this period of privation and uncertainty? The Shuar sold the colonists small amounts of food. CREA periodically contracted with the colonists for the construction of trails with *empalisadas*, logs laid side by side, which lifted travelers out of the mud and improved travel times (see figure 5.4). The contracts provided the colonists with an occasional source of income until their farms began to produce harvests. The advance of the penetration road lifted everyone's spirits. Each colonist returning from a trip to the Sierra brought news of the advancing road. Construction crews completed the Limon to Mendez segment and pushed the head of the road north from Macas toward the cable crossing during the early 1970s. Advances in the road reduced the length of the walk in subsequent trips to the Sierra and brought the colonists' landholdings closer to urban markets. The increase in accessibility meant, as every colonist knew, that the value of their lands would increase. At the very worst

Figure 5.4. A Jungle Trail with an *Empalisada* (photo by M. Gildesgame)

they could, in the future, sell their landholdings for appreciable sums of money.

On occasion politicians, in particular the director of CREA's colonization program, renewed the colonists' commitment with stirring speeches about "building the nation through colonization and community development."[9] A colonist's description of the festivities celebrating the first year of San Carlos de Zamora conveys the spirit of these occasions.

> Just one year ago we began to clear the jungle and incorporate these lands into our national life. On September 15, 1969, with incredible jubilation, we celebrated the first anniversary of the arrival of the brave San Fernando colonists on the banks of the Zamora River. A spirit of brotherly love reigned in San Carlos. Some sixty Shuar from nearby villages came as ambassadors of good will to make toasts and participate in sporting events. Mr. Joel Riley, Mr. William Dewey, the subdirector of the Peace Corps, and Mr. Augusto Abad, the [CREA] director of colonization, graced us with their presence. Together with us were thirty persons from the neighboring cooperative of Santiago de Gualaceo as well as residents from the nearby villages of San Luis, Bomboa, San Jacinto, Sutzo, Panantza, and Catazho. Altogether 250 persons attended the festivities. The program of events was all the more beautiful because for the first time in this immense, shadowy jungle we raised the Ecuadorian flag and sang the sacred notes of our national anthem.[10]

A plaque on the Pan to Mendez mule trail offers more evidence of the patriotic spirit that animates participants in pioneering efforts.

> The municipality of Paute commends and devotes itself to the memory of Manuel Alarcon, Benjamin Flores, Jose Antonio Maldonado, Amadeo Lopez, Francisco Lopez, and Redentor Cardenas. Sons of this parish who died in Ruyaloma on the 11th of July 1917, during the construction of the Pan-Mendez trail. In gratitude and in recognition of Ecuador.
>
> *February 27 (Day of Patriotism), 1959*[11]

Fears as well as hopes played a role in individual decisions to colonize. The history of one family, who persisted in trying to establish a farm in the forest, illustrates the mix of ideas and emotions that sustained colonists in their quest.

Pablo C and Family: A History

In 1968 Pablo C., 33 years old, had a wife, three children, and few economic prospects in the Sierra. He had worked as a wage laborer on coastal plantations and lost several crops to drought on his small Sierra landholding. When the Peace Corps volunteer in San Fernando established a cooperative for colonizing the Oriente, Pablo's wife urged him to "run and sign up" while the cooperative still had openings. Pablo's earlier experiences working in tropical agriculture on the coast shaped his perceptions about the economic possibilities of farming in the Oriente. He planned to plant cacao, a cash crop on the coast, and earn large profits. Settled in San Carlos de Zamora, Pablo and his family struggled to establish their farm, but the continuing isolation of the community and the increasing incidence of malaria eventually persuaded them to sell their land to one of the few remaining families in the area.

Instead of returning to the Sierra, Pablo and his family moved to the Upano—Palora region and settled in La Quinta cooperative where they received 30 hectares of land. He and his sons later acquired an additional 25 hectares of undeveloped land near the Parque Nacional de Sangay. In the early 1980s Pablo fought publicly with several other Quinta colonists about plans to invade the properties of other landowners in the area. In the aftermath of the conflict Pablo moved six kilometers to Sinai where he now lives with his wife and seven children. He has cleared about one-half of his lands and planted pasture, but many of his cattle have died, so he earns little from the land. Most of his income comes from working as a day laborer for Duran. After twenty years in the Oriente and little economic progress, Pablo remains hopeful. Over half of his land remains forested, and he plans to expand the size of his pastures and herd with cattle from the Sierra and help from one of his sons.[12]

A GROWTH COALITION: CREA, DURAN, AND THE COLONISTS

Colonists like Pablo C. showed fortitude and resilience in their efforts to colonize in the rain forest, but they rarely tried to start a farm without assistance. Even Duran, with substantial amounts of capital, needed help. To clear a large expanse of rain forest with axes, machetes, and chain saws, Duran needed a large number of workers. The remoteness of his ranch made it virtually impossible to recruit workers from the surrounding area, so Duran relied on his partners in Cuenca to supply workers when he needed them. Many peasants would not work in such a remote setting, and others came for reasons that raised doubts about their willingness to work. One foreman on Duran's ranch was a fugitive from justice. He had killed a man in a barroom brawl in the Sierra. To secure a supply of reliable laborers, Duran encouraged his steady workers to set down roots in the region by using their days off to develop their own farms on adjoining tracts of land.

The CREA colonists represented an additional supply of labor to Duran, so he decided to help them when they first arrived in the jungle. The colonists could not have established the settlement at Sinai without the access provided by Duran's cable across the Upano. Duran prevented the disintegration of the group when the colonists could not locate an accessible tract of unclaimed land. He guided them to the land they later claimed and cleared. For the first ten years of Sinai's existence, the colonists reached the settlement via a trail cut by Duran's workers that zigzagged up the cliffs on the east side of the Upano. A big, strapping man with a fondness for conversation, Duran provided many a passing colonist with a drink and free meal on their way up the trail from the Upano. On one occasion the rough trails up from the Upano induced labor in the wife of a young colonist, and she gave birth in a cattle pasture on Duran's lands. The mother and the baby convalesced for several days in one of Duran's bunkhouses before they moved on to the recently established town.

Both Duran and the colonists benefited from the arrangements they made to clear land on his ranch. The colonists provided the *hacendado* with a trustworthy source of cheap labor. Duran knew each of the colonists, so he could estimate their ability as workers before he hired them. In contrast he hired his recruits from the Sierra sight unseen. The contracts with Duran provided the colonists with a steady source of income close to home. The colonists' circumstances immediately after settlement made these arrangements important. Their farms did not produce anything of value, so they had to work as wage laborers. A nearby place of work reduced the time and money spent in transit and enabled the colonists to live at home and work intermittently on their own lands while earning money to support their families.

Relations between the colonists and Duran changed dramatically when a representative of IERAC, the agrarian reform agency, arrived in Sinai to measure land early in 1970. The IERAC official asserted that, because Duran had more land than the law allowed, IERAC would expropriate his lands and award them to the colonists. When the official recruited eight colonists for a surveying team and began to subdivide Duran's land, a confrontation occurred. In Duran's words, "the IERAC officials and the colonists began to cut boundary lines across my property. I met them in the forest and said, 'I don't want to fight. I have come alone, and you have eight men with machetes. But I am going to tell you one thing. You are opening up boundary lines on my property. These boundaries will serve me well because I am going to come in behind you with fifty workers clearing land and wiping out your boundary lines.'"

In addition Duran cut off the colonists' access to the cable over the Upano and lobbied at IERAC's headquarters in Quito for a change in policy. The last strategy paid quick dividends. Under pressure from Duran's partners in Cuenca, the head of IERAC transferred his representative in Sinai to a post 400 kilometers away in coastal Ecuador. Without the IERAC official to provide leadership, the colonists abandoned the invasion, and Duran reopened the cable crossing over the Upano. The hostility generated by the invasion gradually dissipated. Five years after the invasion Duran began renting pasture to one of the eight invaders.

A variable pattern of mutual aid and class conflict characterized relations between the colonists and the other large landowner in the region, J. Monsalve. Unlike Duran, Monsalve was an absentee owner. During the initial period of settlement he never visited his land, now he visits the land once or twice a year. A series of managers living on the property have developed the farm, perhaps 900 hectares in extent, according to Monsalve's directives. Like Duran, Monsalve provided a convenient source of work for the colonists from Sinai.

He also endured a land invasion in the mid-1970s when several families from La Quinta community claimed the forested portions of Monsalve's property in the name of a bogus cooperative they had formed. Again, the large landowner's political influence in the capital proved decisive in thwarting the invasion. Here too, the relationship between the large landowner and the surrounding smallholders proved to be both fluid and varied. After the local authorities refused to support the invasion, the colonists withdrew and turned to other pursuits. Those colonists who did not join the invaders continued to work as contractors for Monsalve, clearing 5- to 10-hectare tracts of forest for him.

Of the several coalitions active on the Upano—Palora plain in the early 1970s, the one with the most impact on the landscape involved CREA, colonists, and large landowners. Each group contributed to and benefited from the other groups' activities. CREA constructed the road, which benefited both the colonists and the large landowners from Cuenca who developed ranches in the path of the road.[13] CREA also benefited. By settling colonists from Azuay in the area, CREA complied with the conditions of their IDB loan and enlarged Cuenca's hinterland, thereby contributing to the prosperity of the city's economic elite who staffed the agency. The large landowners helped the colonists, and the colonists provided the landowners with a steady supply of cheap labor. In effect the colonists and the large landowners helped each other deforest the region, and both groups profited from the increased land values that accompanied the deforestation.

Like other coalitions, this tripartite coalition involved regional as well as local elites, and it brought together people from several different economic classes. The diverse interests of the coalition's members underline the potential fragility of these alliances. The CREA / colonist / *hacendado* coalition disintegrated during the early 1970s when conflicts broke out between the colonists and the large landowners. The short history of this coalition suggests that many coalitions have fleeting existences. Despite their short lives, they have a lasting impact on regional landscapes because they provide the infrastructure that spurs deforestation.

The sequence of initial peace, followed by intense conflict, and then a return to peace in the Upano—Palora region suggests that the incidence of land conflicts varies over time and place. The conflicts cluster in places where the construction of roads and other infrastructure induces rapid rises in land values. By the late 1980s aggressive land speculators had pretty much abandoned the Upano—Palora region. Only the area near the recently completed road and ferry crossing of the Pastaza River continued to experience extensive speculation in land.[14] In 1990 Shuar living near the ferry crossing accused the colonists of redrawing some boundaries, and conflict erupted between the two groups. The conflict did not spread to other communities in the Upano—Palora region.[15] Colonists in the Morona region of southeastern Morona Santiago did not experience any conflicts over land until 1986 when the pending completion of the Mendez–Morona highway attracted speculators interested in acquiring large tracts of land, and land prices began to increase.[16]

While rapid increases in land values contribute to conflicts over landownership, subsequent periods of stability in land markets after a road's arrival reduce the number of conflicts over land. These changes in the incidence of conflict affected the rate of deforestation in the Upano—Palora region because colonists frequently cleared land in anxious attempts to secure title to it. By the mid-1970s the Shuar, the colonists, and large landholders such as Duran had cleared so much land to establish ownership that they had more pasture than their cattle could use (Federacion 1976:267).[17] In the more tranquil atmosphere that followed this acquisitive fever, no one had legal titles to land, but informal understandings about the location of boundaries developed between neighboring landowners. The same type of consensual understanding also made it possible to buy and sell land in the Upano—Palora region without titles. People knew that you owned a tract of land, and they knew the person to whom you sold the land. With land plentiful and cattle scarce, incentives to grab land and hold it for speculative profits declined. Under these circumstances colonists could leave a tract of land in forest without worrying that the land would become the target of an invasion because it had not been worked. In this changed atmosphere a new set of dynamics, closely linked to variations in household economics, began to influence the rate of deforestation.

6

Clearing Forest Remnants: The Upano—Palora Plain in the 1980s

Fifteen years after Duran and the colonists crossed the Upano and began to clear land, only fragments of the original forest remained in the region (see figure 6.1). People continued to clear land, but they did so under quite different circumstances than the first settlers. The altered circumstances should, according to the argument advanced in chapter 2, change the causes of deforestation. Organizations which directly or indirectly support pioneering smallholders should decline in importance, and factors like numbers of children and access to credit should assume much more importance in explaining the continued clearing of land. The following analysis of deforestation on the Upano—Palora plain during the 1980s provides an opportunity to assess this claim about changes in the causes of deforestation. The chapter begins with descriptions of how the region's first road changed the landscape, and it concludes with an analysis of household differences in land clearing which explains the emerging pattern of land use.

THE ROAD AND DEFORESTATION

Between 1965 and 1983 colonists and Shuar cleared almost half of the land on the Upano—Palora plain (see table 4.3). The construction of a road north from Macas during this period changed the local agricultural economy and stimulated extensive land clearing. Before the government built the road, landowners earned all of their income from cattle. Itinerant merchants went from farm to farm, purchasing cattle for low prices and driving them north to Puyo for shipment to the Sierra. Access to the new road made it possible to transport crops like coffee (*C. arabica*) and narangilla (*Solanum quitoense*) to urban markets for the first time. When the road reached Santa Rosa, a five-hour walk south of Sinai, the colonists began planting narangilla; twice a week they would pack the narangilla fruit into boxes, load the boxes onto mules, and make the trip to Santa Rosa to sell the fruit to truckers. Because narangilla produces best on well-rested soils, colonists cleared primary forest to plant it. When the yields from these plantings began to decline after several years, the colonists converted the narangilla groves into pasture. During the first few growing seasons the isolation of the new fields protected the crops from pests, and the colonists earned high profits, as much as US$900 per hectare in good years. By 1978 most of the colonists in Sinai had begun clearing a hectare of primary forest each year for the cultivation of narangilla.[1] In the early 1980s the price of the fruit fluctuated widely and problems with pests increased, but the colonists persisted with the sequence of planting narangilla first and sowing pasture several years later.

Shuar land use also changed with the arrival of the road. Like some colonists, the Shuar began selling the valuable wood on their property to loggers. Shuar and colonist landowners sold all of the laurels (*Cordia alliodora*) on tracts of land up to six kilometers from the road. Working on felled trees with two-man saws, the loggers would cut boards out of the logs and haul the boards out to the road with mules (see figure 6.2). In some instances the Shuar cleared land close to the

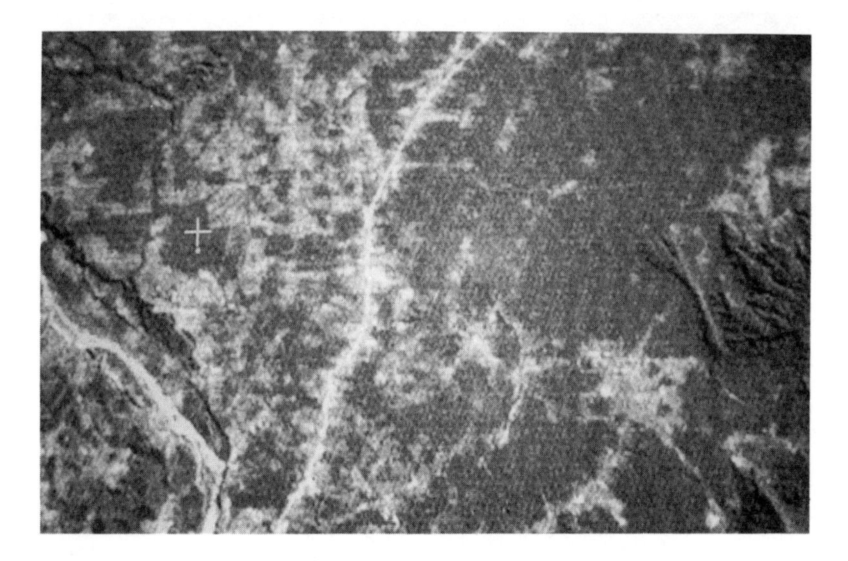

Figure 6.1. Land Cover on the Upano–Palora Plain, 1983 (source: LANDSAT)

road and planted pasture with the intention of renting it out to nearby colonists or passing merchants.

While the improvements in transportation brought a measure of prosperity to the early colonists, continued difficulties with access to the region limited agricultural expansion. River crossings kept marketing costs high. In the first fifteen years after the construction of the first bridge over the Upano at Macas, shifting currents and flimsy construction caused the bridges to collapse three times (see figure 6.3). Each time a collapse occurred, it took months to repair the bridge, and in the interim the colonists and the Shuar paid increased freight charges.

Under these circumstances rather than paying one freight charge from Sinai to Cuenca, farmers paid three charges: one for the transport of goods from Sinai to the river crossing, another for the river crossing where porters carried goods across a footbridge, and a third charge for transporting goods from Macas to Cuenca. In order to make a profit under these conditions, middlemen lowered the prices they paid for agricultural commodities at the farm gate. Another haz-

Figure 6.2. Horses Hauling Boards Into Sinai (photo by T. Rudel)

ardous river crossing prevented the colonists from selling goods in markets to the north. Under low water conditions several mestizos operated a makeshift ferry on the Pastaza River. It consisted of a platform mounted on two canoes powered by outboard motors. When the river began to rise after a rain, the ferry's owners would dismantle it to prevent it from disintegrating in the turbulent waters of the swollen river. The intermittent ferry service made it impossible to market perishable goods like narangilla in the urban markets of the northern Sierra. These difficulties lowered the profits which Upano–Palora farmers could expect to earn and discouraged them from clearing additional land.

THE GEOGRAPHY OF DEFORESTATION

Most Upano—Palora smallholders continued to clear land despite the obstacles, but some pushed ahead more vigorously than others. These local differences in deforestation stand out in satellite images of the region during the 1980s. Some areas show extensive deforestation while other areas do not (see figure 6.1). Analyses of a 1983 satellite

Figure 6.3. A Collapsed Bridge Across the Upano River (photo by T. Rudel)

image indicate both the extent and the possible origins of these differences. A detailed map, which indicated boundaries between Shuar and colonist lands, made it possible to identify the boundaries of twenty-one Shuar and colonist communities on the 1983 satellite image.[2] With the boundaries identified, we could measure the extent of forest cover in ten colonist and eleven Shuar communities in the southern portions of the Upano—Palora plain.[3] A statistical analysis of these data explains the extent of forest cover in terms of a community's location, its age, and the ethnicity of its residents.[4]

The correlation matrix in table 6.1 indicates some expected patterns of land clearing. New communities had more forest than old communities (r=.49), and places farther from the road had more forest than places closer to the road (r=.48). Among mestizos, smallholders had cleared a much larger proportion of their land (76%) than large landowners (38%).[5] Large landholdings contain the largest remaining patches of forest. The cross in figure 6.1 denotes a large forested area on Duran's hacienda, the largest landholding in the region. Forest cover appears to follow the familiar *latifundia-mini-*

fundia pattern in which large landowners leave extensive areas in fallow while smallholders work the land intensively.

While figure 6.1 shows the faint outlines of an agricultural corridor along the road, this pattern is more pronounced in colonist zones than it is in Shuar zones. Proximity to a road induces the colonists to clear more land (r=.51), but it has little effect on land use among the Shuar (r=.16). Three Shuar *centros* have relocated their village centers to the road, and a number of individuals have established roadside homes, but in most places the extent of cleared land varies more with the internal organization of a *centro* than with proximity to the road. The strong market orientation of the colonists accounts for the concentration of their fields along roads. The colonists not only cleared a greater proportion of their land along roads, they also owned most of the land close to the roads. This pattern of ownership reflects the acquisition strategies of the colonists, purchasing lands in the path of roads under construction or along trails that later became roads. Not surprisingly, ethnicity associates strongly with deforestation. Although the colonists and the Shuar founded their communities at approximately the same time, the colonists cleared a much larger proportion of their land (66%) than the Shuar (29%) during the following twenty years. The regression analysis, presented in table 6.1 confirms these findings.[6] Ethnicity and the length of time since the founding of a community explain most of the intercommunity variation in the extent of deforestation.

INDIVIDUAL DIFFERENCES IN LAND CLEARING

To understand why ethnicity has such a marked effect on deforestation rates, we interviewed colonist and Shuar families in two communities on the Upano—Palora plain. The two communities resemble one another in several crucial respects. First, the inhabitants began to clear land about the same time, during the 1960s. The Shuar living in the Uunt Chiwias region wanted to raise cattle on individual farms, so in 1966 they asked the Salesians to establish boundaries for the community and its interior lots. Three years later the colonists settled in Sinai. Second, the topography, soils, and climate of the two places are similar. They are located about twenty kilometers apart on the plain

Table 6.1: Land Use in Twenty-One Upano—Palora Communities, 1983

Panel A: Land Use by the Ethnicity of the Community

	Ethnicity	
Variable	Shuar	Colonist
% land cleared	29.5	66.6
Distance from road	7.9	2.1
Date founded	1963	1961

Panel B: Correlation Matrix

	(1)	(2)	(3)	(4)
(1) % land forested				
(2) Distance from road	.48*			
(4) Date founded	.49*	.21		
(5) Ethnicity	-.73***	-.44*	-.13	
Mean	52.8	5.1	1962	1.47
Standard Deviation	25.8	6.7	7.3	51

Panel C: Regression Analysis

% Forested = 25139 - 307 Ethnic*** + 13 Date* + 5.4 Distance
(69) (3.8) (4.7)

r = .88, r2 (adj) = .73, p = .000, n = 21

N.B. *** <p=.001, **<p=.01, *<p=.05. The numbers preceding the variable name in Panel C are regression coefficients; the numbers in parentheses are the standard errors for these coefficients.

between the Upano and Palora rivers (see figure 6.4). Finally, the same network of roads traverses both communities, so access to markets does not vary much between the two places.

Households are the units of analysis.[7] To get information about land use and household economics, we interviewed all the landowners in Uunt Chiwais and all of the landowners in the area originally settled by the Sinai colonists.[8] These procedures resulted in thirty-three usable interviews from Uunt Chiwias and thirty interviews from Sinai. The analysis of variations in the extent of land clearing employs four variables: cleared land, ethnicity, land area, and household size. The measures of all four variables are straightforward.

The dependent variable is:

cleared land: the proportion of land owned by a household head that has been cleared.9

The independent variables are:

ethnicity of household head: either Shuar or mestizo.
household size: the number of family members living with the respondent.
land area: the amount of land owned by the household head.

Table 6.2 presents the results of the survey. The data reveal a pattern of land use consistent with the patterns in the larger region. Although the colonists and the Shuar settled their lands at the same time and have similar-sized tracts of land, the colonists have cleared twice as much of their land (76%) as the Shuar (38%).

In the regression analysis presented in panel B of table 6.3, all three independent variables in equation one explain some of the variation in land clearing. Households with smaller tracts of land clear a larger proportion of their lands. Larger households clear more of their land than do smaller households. Columns 4 and 5 of Panel B present regression analyses of the colonist and Shuar subsamples. The only notable difference involves the significance of household size for land clearing. Given the almost complete integration of the colonists into the market economy and the land extensive nature of cattle ranching, even the smallest colonist households use almost all of the land available to them on smallholdings, so household size does not account for much deforestation on small colonist landholdings. The largest colonist landholdings, above 500 hectares, have proportionally more forest than small colonist landholdings. This finding from the remote sensing analysis, combined with the results from interviews with colonist smallholders, leaves open the possibility that household size begins to affect colonist land clearing only when they have more than 60 or 70 hectares of land. Among the Shuar, with their more pronounced subsistence orientation, household size has a marked effect on land clearing, even on small farms. Each additional person in the household prompts more land clearing.

Ethnicity predicts the extent of cleared land on a farm. As indicated in table 6.2, the colonists use commercial credit more than the Shuar, 90 percent to 25 percent, and these differences in the use of credit could account for the ethnic differences in deforestation. Equations (1), (2), and (3) in table 6.3 explore this possibility with a dummy variable for the use of commercial credit. The credit variable does not add anything to an equation (#3) which contains the land,

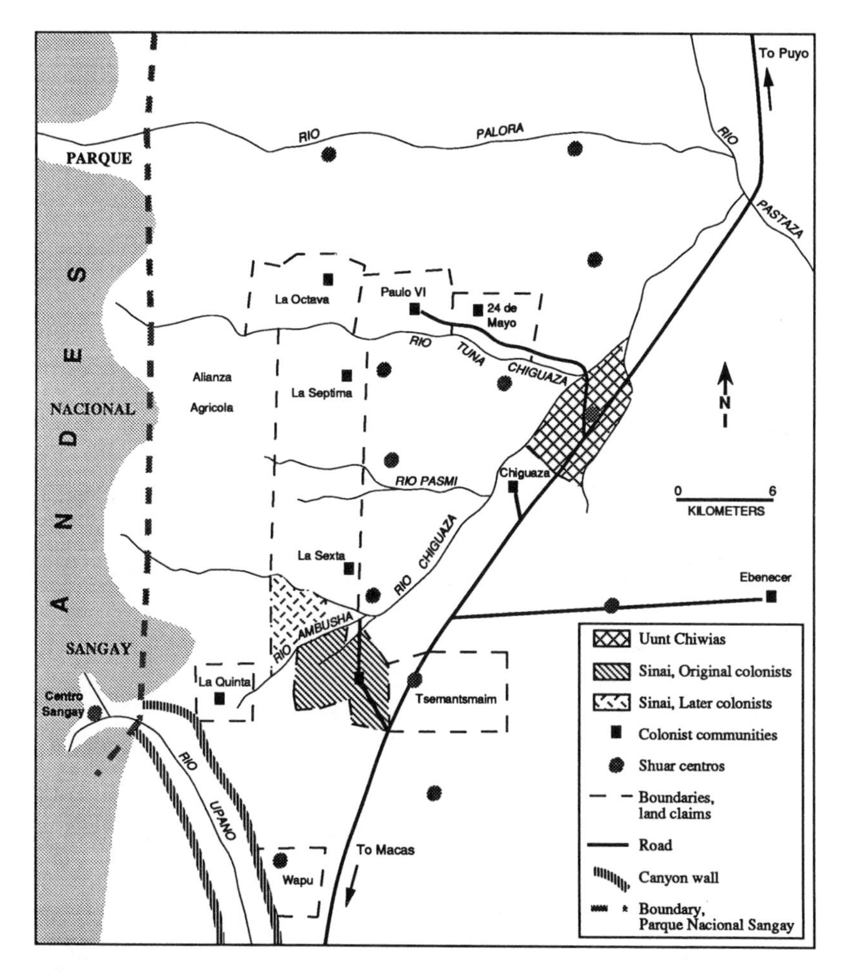

Figure 6.4. Settlements on the Upano---Palora Plain

Table 6.2. Deforestation Among Landowners in Sinai and Uunt Chiwias

Attributes of Smallholders in the Two Communities

	Means	
Variables	Colonists	Shuar
Land cleared	76%	38%
Household size	6.8	5.9
Land (in hectares)	61	63
Cattle	24	7
Cattle/Pasture	.602	.336
Households with bank loans	90%	25%

household, and ethnicity variables. Equation (2) substitutes the credit variable for the ethnicity variable, and in combination with equation (1) it provides a measure of the relative significance of the two variables. The differential use of credit explains some of the variation in the extent of deforestation, but the decline in variance explained from .53 in equation (1) to .24 in equation (2) indicates that ethnicity explains more variation in deforestation than does the use of credit. These findings suggest that other ethnic differences besides the use of credit figure centrally in the greater willingness of the colonists to clear land. The following sections examine the use of credit and other ethnic differences that have contributed to the different patterns of land clearing among the colonists and the Shuar.

CREDIT

Shortages of pasture grass and cattle, not credit, set limits on land clearing during the first few years of settlement. The most commonly used pasture grass, gramalote (*Panicum purapurascens*), reproduces vegetatively, so colonists had to transport shoots of the grass to the new fields. To minimize the loads of grass they had to carry, the colonists planted the grass sparsely, as much as four meters between stalks. In many cases the tangle of secondary growth that sprouted in these fields could not support cattle, so the colonists had to replant the fields. Even with some pastures unable to support cattle, the Shuar,

Table 6.3 Regression Analyses of Deforestation in Sinai and Uunt Chiwias

Panel A: Correlation Matrix

	(1)	(2)	(3)	(4)	(5)
(1) Land cleared					
(2) Land	-.150				
(3) Household size	.260*	.418***			
(4) Ethnicity	-.692***	.028	-.158		
(5) Bank loans	-.408**	.173	.298*	-.562***	
Mean	56.2%	62.2	6.37	1.51	.673
Standard Deviation	20.0	35.2	3.05	.50	.489

Panel B: Regression Analyses

Equation #:	Eq. (1)	Eq. (2)	Eq. (3)	Eq. (4) Colonists	Eq. (5) Shuar
Variable					
Size of landholding	-.186*	-.257**	-.187*	-.240*	-.185*
	(.075)	(.095)	(.076)	(.101)	(.095)
Household size	.023**	.025*	.023**	.003	.045***
	(.009)	(.011)	(.009)	(.011)	(.011)
Ethnicity	-346***		.343***		
	(.048)		(.058)		
Bank loans		.210**	.007		
		(.065)	(.062)		
r=	.743	.529	.743	.455	.591
r^2 (adj.) =	.528	.242	.520	.148	.302
p=	.001	.001	.001	.04	.002
n=	63	63	63	30	33

N.B: p<.001 = ***, p<.01 = **, p<.05 = *. In Panel B the top numbers are regression coefficients; the numbers in parentheses under the regression coefficients are their standard errors.

the colonists, and the large landowners still found themselves clearing land faster than they could acquire cattle. Few landholders, Shuar or colonist, had cattle for sale in the region, and the new landowners did not want to purchase cattle in distant regions because the rigors of the long trip into the region, with rough trails and river crossings, killed many cattle.

When the shortages of pasture grass and cattle eased several years after settlement, the colonists and the Shuar began to look for both formal and informal sources of credit to finance land clearing and cattle acquisition. Both groups acquired cattle through informal arrangements, but the colonists used formal sources of credit more frequently than the Shuar, and overall the colonists used more credit than the Shuar. Differences in access to credit explain in part why the colonists used more credit than the Shuar. The Banco Nacional de Fomento (BNF) lent money to the colonists, and the Federacion de Centros Shuar made loans to the Shuar.

Credit for Colonists

Despite their remote location and skimpy household budgets, the colonists acquired over five hundred head of cattle during their first five years in Sinai.[10] The first cattle arrived in Sinai in 1970 after several colonists entered into *a medias* agreements with partners from the Sierra who provided the cattle or funds to purchase them. The colonist would pasture the animals on his lands and keep every other offspring from the herd. With an undeveloped tract of land in a remote setting as their only collateral, the poorer colonists could not qualify for a loan from a bank. They received loans from the BNF in 1971 only because CREA cosigned the loans, agreeing to supervise the collection of payments and to repay the loans if the colonists defaulted.[11]

Impressed by the large sums of money entrusted to them, some of the young colonists received their first loans "with trembling hands." The loans accelerated the rate of land clearing. Before the colonists received the loans, they usually alternated periods of wage labor with work on their farms. After they received the loans, the colonists used the principal for living expenses and devoted all of their time to land clearing. Under these conditions deforestation rates jumped. In a typical case a colonist cleared approximately 2 hectares of land a year until 1974 when he received a loan. Over the next two years he cleared 16 hectares of land and then returned to his previous pattern of clearing several hectares a year. With an infusion of government funds from oil exports, the BNF kept the line of credit for land clearing open until 1975.[12]

After the initial loans for land clearing, the colonists began taking out loans to purchase cattle. Easy credit for cattle encouraged land clearing. When smallholders expected to receive a loan to expand the size of their herds, they would clear additional lands to support the larger herds. The BNF's terms on cattle loans favored borrowers. The rates on these loans remained fixed at 6 percent throughout the 1970s even though the rate of inflation averaged 11.9 percent for the decade (Yaccino 1988; Wilkie and Haber 1983). Because the colonists repaid the loans in depreciated currency, the government, through the BNF, in effect subsidized the purchase of cattle by the colonists. Despite the favorable terms, some colonists found it difficult to repay their loans.

When the first cattle arrived on a farm, work routines changed because someone had to move the cattle to new pastures each day. The change in routines created problems for families with small children. The men had previously left home for weeks at a time to work as contract laborers on ranches and plantations. Because women with small children could not care for the cattle, the men could not leave now unless they hired someone to care for the cattle. The cost of the additional labor ate into the wages which the colonists earned to support their families.

The young families solved this problem with the BNF money. Because the BNF loaned them a generous amount of money per cow, many families could live off a portion of the loan and still purchase the number of cattle stipulated in the loan agreement. With less money to purchase cattle, these families often purchased young or sickly cattle. Disproportionate numbers of these cows died by strangulation or through overexertion in the rough, timber strewn pastures around Sinai.[13] When the cattle died, these families lost the means to pay off the loans to the bank.

Large families and families with older children did not face this problem. They did not have to forego off-farm labor to care for their cattle, so they did not have to use BNF money for subsistence and could use all of the loan money to purchase cattle. They bought better cattle, paid off their loans, and eventually began to purchase additional cattle with their own money. In contrast many of the families with younger children had to sell off their remaining cattle after a few years in order to make payments on the loans. At best these families made enough payments on their first loan to qualify for a second loan; in their words they "lived from loan to loan."[14] Their herds did not increase, and they cleared less land than their more prosperous neighbors.

Small households without children had more flexibility in sources and amounts of income. They could withdraw entirely from cattle ranching and still survive. For example, when a bachelor purchased cattle with a bank loan and some of them died, he faced the prospect of losing his land because he could not repay the loan. He sold the remainder of his cattle and repaid the loan with revenues from the sale of wood on his lot. He survives financially on a small salary as a border guard for Sangay National Park. The added responsibilities of a family with children would have forced him to chose the riskier option of cattle ranching with its potential of a higher household income.

The same pattern of household decision making characterizes land use in the foothills bordering the national park, a three- to four-hour walk west of Sinai. According to a park ranger, the sons of Sinai's original settlers "grabbed the land and then went off to school, abandoning their new farms for long stretches of time." Only the young heads of household with numerous children persisted in clearing land in these isolated locations. Presumably the need for an increasing flow of income accounts for the persistence of the young fathers in developing their farms. The high ratios of dependents to workers in these households made it more difficult to build cattle herds and increased the pressure on heads of households to exploit marginal economic opportunities.

Credit for the Shuar

The Shuar in Uunt Chiwias got their first introduction to cattle ranching during the 1940s and 1950s when they worked as wage laborers on mestizo-owned cattle ranches around Macas. These arrangements taught the Shuar the value of cattle in a gruesome way. If a cow died while under the care of a Shuar worker, he had to repay the owner the value of the cow in labor. This penalty often meant six months of unpaid labor for the Shuar.[15] They came to regard the ownership of cattle as a valuable way of storing wealth for use in emergencies, so they began purchasing cattle in Macas and driving them to outlying *centros* in the early 1960s. Missionaries also brought cattle into the *centros*. The Salesians established herds at their missions to provide milk for the children at the mission schools. Once the missions' herds reached a large enough size to provide for the children's needs, the priests began lending their cattle *a medias* to the Shuar.

In the early 1970s the Federacion de Centros Shuar began to extend loans to cattle development groups in the *centros*. As noted earlier, the

Shuar in areas experiencing heavy colonist pressure had decided to clear large portions of land and become cattle ranchers in order to retain control over the land.[16] The Shuar's initial plans for cattle ranching progressed slowly because the BNF would not loan them money for the purchase of cattle unless they had title to their land, and many *centros* did not have titles. Although some *centros* received titles in the late 1960s, others did not. As late as 1988 only 32 percent of all Shuar *centros* had received titles to their land (Fundacion Natura 1988). For these communities the Shuar program offered the only available source of credit for intensifying agricultural production. Villages with titles could obtain loans from the BNF (Federacion 1976:153), but, if a *centro* failed to repay a loan, as occurred in the late 1960s near Macas, the mestizo-controlled BNF could seize the *centro*'s land. Just as a regular source of income seemed an oddity, so regular payments to a bank seemed an intolerable burden to many Shuar. The leaders of the federation feared that many Shuar would lose control of their lands through defaults on loans from the BNF. To prevent the foreclosures, the federation established its own loan program and advised against taking out loans from the BNF. Under this new program the Shuar would at least retain control over the land when villages failed to repay their loans.

The federation began its credit program with funds from German, Dutch, and American foundations. To receive a loan, individuals in a *centro* would form a "cattle development group" which planted pastures on individual plots of land and applied for loans to purchase cattle. The individuals then divided up the purchased cattle among themselves, but they retained a collective responsibility to pay off the loan. Like the BNF's loans, the federation's loans stipulated an interest rate (10%) below the prevailing inflation rate, so, like the BNF, the federation subsidized the purchase of cattle.

During its first five years, the program made cattle loans to approximately one half of the *centros*. The villages west of the Kutucu experienced more pressure from encroaching colonists and had easier access to markets, so they took out most of the loans. The western *centros* comprised 66 percent of all *centros* during the mid-1970s, and they took out 84 percent of the loans during the first five years of the program.[17] Pressures from colonists and access to markets also affected cattle acquisition east of the Kutucu. Of the eleven loan recipients east of the Kutucu, four went to Shuar communities clustered around Taisha, a large Salesian mission, where colonists had began to compete with the Shuar for land during the 1970s.

Historical differences in the internal organization of *centros* affected their ability to get loans and clear land. Some *centros* developed around cohesive groups; Shuar from the same Upano valley village moved to an uninhabited area east of the Kutucu and formed a village there. Other villages formed around the extended families of celebrated shamans or "big" men (Hendricks 1986:20–30). Some *centros* had artificial origins. To secure a tract of land, federation officials convinced unrelated Shuar families living in the same area to form a *centro*. The residents of artificial *centros* may have had more difficulty forming cattle development groups than the residents of other *centros*, and the corresponding lack of credit slowed the rate of deforestation in these places.

The cattle cooperatives varied considerably in their levels of activity across communities, and these variations contributed to some surprising patterns of deforestation. For example, comparing four *centros* located in the Upano—Palora plain, the two villages with strong leadership, located far from roads, had cleared a larger proportion of their land, 27 percent to 19 percent, than the two villages with weak political leadership and locations along roads.[18] The fortunes of cattle development groups also changed over time. In the early 1970s Uunt Chiwias had a "showcase" cooperative; its members cleared land, obtained a loan, and paid it off in the mid-1970s. Despite all this activity the cooperative fell apart in the late 1970s. The loss of a leader often explained why a group disintegrated.[19]

The Shuar, like the colonists, created coalitions to secure land and later used them to acquire cattle. By promoting the conversion of forests into fields, the coalitions acquired lands for their people, and at the same time helped individuals realize economic gains. The increase in pastures and numbers of cattle raised the incomes of individual Shuar, increased the value of their lands, and enabled some of them to realize gains by selling their land and moving elsewhere.[20] Although the Shuar coalitions hastened deforestation in many places, they did not have as dramatic an impact on the landscape as the alliance between BNF, CREA, and the colonists. The Shuar program of credit produced deforestation only when fragile collective organizations prospered in a place; the effects of the alliance between the colonists, CREA, and the bank did not have this intermittent quality. CREA no longer guarantees the colonists' loans from the BNF, but the connection it established between the banks and the colonists persists. Each year a sizable minority of the colonists in Sinai take out loans to expand their cattle herds and clear forested land.

NORMS

Differences in values account for some of the ethnic differences in land use. The original colonists in Sinai took pride in their ability to clear land. They recalled with satisfaction how they cleared 100 hectares for Monsalve or 60 hectares for a landowner north of Macas. For years after the settlement of Sinai colonists measured a man's worth by the amount of time it took him to clear one hectare of land. A strong worker could clear the land in two weeks; weaker workers would take a month or more to clear the same piece of land. Few people complained about the hard physical labor involved in clearing land. Almost all of the colonists were young men when they arrived in Sinai; they anticipated hard work in their first few years in the jungle, followed by material success and less demanding work in their old age. As Don Pasato, one of Sinai's strongest workers put it, "you have to work in your youth in order to sit in old age."

In contrast with the colonists, the Uunt Chiwias Shuar expressed ambivalence, even regret, about the need to clear land. It distressed them to see the forest destroyed. Some Shuar wanted to preserve the forest for specific reasons. They wanted to retain large reserves of forest in order to preserve a population of animals for hunting, or they preferred forests to pastures because a forested landscape is "healthier" than a cleared landscape. This last observation may reflect the impact of increased population densities on the health of the Shuar. In the densely populated Upano valley *centros*, where almost all of the land is cleared, medicinal plants are harder to find. Contaminated water supplies almost certainly make intestinal parasites more prevalent in densely populated *centros* than they are in the sparsely populated, forested *centros* east of the Kutucu range. Finally, the Shuar complained about cattle; they objected to the dirt and the smells around cattle in a pasture. Even the meat failed to find favor. The Shuar east of the Kutucu preferred to eat wild boar.[21]

Differences in attitudes toward the forest may explain some of the differences in the agricultural economies of Shuar and colonist communities. While more than 80 percent of the households in Sinai remain committed to cattle ranching as their primary source of income, only 35 percent of the households in Uunt Chiwias earn most of their income from the sale of cattle, and there are signs of declining enthusiasm for cattle ranching among the Shuar. The colonists maintain much larger herds of cattle than the Shuar, averaging 24 head of cattle per household as opposed to 7 head for Shuar households. Although the Shuar have cleared much less pasture than the colonists, a lack of pasture does not explain the small herds. The Shuar have more pasture than their cattle can use. While the colonists have 1.6 hectares of pasture for every animal, the Shuar have nearly 3 hectares of pasture for every cow. Although we could not confirm it directly in interviews, the Shuar in Uunt Chiwias do rent their pastures to colonists with cattle. Shuar in other western *centros* also rent pastures to colonists.[22]

In the year prior to the survey the Shuar in Uunt Chiwias sold 105 head of cattle and retained a herd of 213 animals. To maintain a herd of constant size in this region a rancher can not sell more than 30 percent of his herd in a single year. In 1986 the Shuar sold approximately 40 percent of their herd. At this rate their herds would decline to insignificant sizes in several years. Among the eight respondents who cited reasons for selling cattle, medical emergencies figured prominently in five cases. Rather than a regular source of income, cattle continue to be a store of wealth for use during crises.

While the extraordinary decline in the number of cattle in Uunt Chiwias may represent nothing more than a short-term fluctuation in the size of a herd, two other pieces of evidence suggest that the decline is part of a long-term trend. First, trends over time in cattle stocking in the Upano—Palora region suggest stability in colonist cattle ranching and change in Shuar cattle ranching. The 1973 stocking rate in Sinai, .59 cows per hectare, is virtually the same as the .60 stocking rate for 1986. Surveys of Shuar *centros* in the Upano—Palora region during the early 1970s reported stocking rates of .49 and .44 cows per hectare.[23] In contrast our 1986 survey in Uunt Chiwias yielded a stocking rate of .33 cows per hectare. The surveys among the Shuar covered overlapping but not identical sets of communities, so they are not directly comparable. Even with this qualification, the results still suggest a declining commitment to cattle ranching.

Shuar plans for their land also suggest a declining interest in cattle ranching. Almost two-thirds of the Uunt Chiwias Shuar already grow narangilla and coffee commercially, and 45 percent of them indicated that they planned to expand the size of their narangilla and coffee plantings. In contrast only 16 percent of the Shuar had plans to expand the size of their cattle herds. The same pattern of increasing reliance on the sale of garden crops characterizes other *centros* in the Upano—Palora region. The Shuar in Wapu have commercial interests in garden crops and small herds of cattle, approximately three cows per family. Their leaders have lobbied for the construction of a feeder road from the main road into the village center. The new road will make it easier for villagers to sell plantains and manioc from their gardens.[24]

In sum the Shuar's commitment to cattle ranching seems half-hearted. Born out of anxieties about control over their ancestral home in the face of colonist invasions, this strategy helped the Shuar retain control over a portion of the lands in western Morona Santiago. Fifteen years later, with access to land assured, most Shuar in the Upano—Palora region appear to be abandoning attempts to make the radical change from shifting cultivation to cattle ranching. They prefer the more incremental change from shifting cultivation to cash cropping.

THE CONVERSION TO CASH CROPPING

A series of factors have persuaded many Shuar and some colonists to cultivate more cash crops. First, the Shuar have traditionally regarded work in their gardens as extremely important. In the words of one Uunt Chiwias Shuar "the gardens are for survival; everything else is less important." Given the importance attached to garden work, its expansion does not require a drastic change in Shuar priorities. Second, the Shuar have a fund of accumulated knowledge about narangilla and coffee because for decades they have cultivated both plants in their gardens. They can use this knowledge if they decide to expand the size of their coffee and narangilla plantings. In contrast the Shuar know very little about cattle. The successful care of a herd requires that they absorb and put to use an entirely new body of knowledge about veterinary medicine. Third, narangilla and coffee production are labor but not capital intensive enterprises, and these production requirements suit the Shuar well. Family members can meet the sporadic but heavy demands for labor in coffee and narangilla cultivation because they usually do not have regular jobs which would complicate participation in a harvest. The low capital requirements of coffee reduce the need for bank loans, so clouded land titles and the corresponding lack of access to banks do not handicap the Shuar. Fourth, the shift away from cattle ranching and into cash cropping postpones the emerging problem of land scarcity among the Uunt Chiwias Shuar. The second- and third-generation Shuar who have received 15-hectare plots of land cannot cultivate gardens and maintain a herd of cattle on such small tracts of land. Even allowing for the fallowing of land, a Shuar could maintain a garden and grow significant amounts of narangilla and coffee on a 15-hectare tract of land. Individuals with larger tracts of land could leave a large proportion of their land in forest.

With their growing reliance on cash cropping, the acculturated Shuar of western Morona Santiago now resemble peasants more than subsistence producers. They produce partially for household consumption and partially for sale in markets. Like most peasants, they "constitute part societies with part cultures. They are definitely rural; yet [they] live in relation to market towns. . . . They lack the isolation, the political autonomy, and the self-sufficiency of tribal populations, but their local units retain much of their . . . identity, integration, and attachment to [the] soil" (Kroeber 1949:284).

Cash cropping has also increased in colonist communities, but it has done so without a corresponding decline in cattle ranching. Two surveys of narangilla cultivation in Sinai, one in 1978 and the other in 1986, indicate a trend toward cash cropping among the colonists. In 1978 the colonists had 38 hectares of narangilla under cultivation; in 1986 a 50 percent sample of villagers had 56 hectares under cultivation.[25] A conservative extrapolation from the 1986 sample to the village population suggests that the land area devoted to narangilla production doubled between 1978 and 1986. Undoubtedly, the increased access to urban markets provided by the penetration road encouraged the cultivation of perishable crops like narangilla among both the colonists and the Shuar. The colonists' growing recognition that they can profit by cultivating narangilla for several years on lands they eventually plan to use as pasture has contributed to the increase in cash cropping. Several producers have abandoned cattle ranching and specialized in the production of narangilla despite the dramatic fluctuations in its price. These landowners have adopted an intensive form of shifting cultivation in which they grow narangilla for two years, let the land lie fallow for seven years, and then replant narangilla. Other colonists pursue a double-barreled strategy; they have begun to devote more land to cash crops without abandoning cattle ranching. The additional land for the cash crops comes from the few remaining tracts of rain forest in the community.

The size of the extended family figures centrally in many decisions to convert forests into fields because the cultivation of most cash crops requires abundant supplies of labor. When the rising price of narangilla in the mid-1980s prompted landowners to clear land and plant narangilla, the most frequent arrangements involved older brothers with land and their younger siblings without land. The older brothers supplied the land; the younger brothers provided the labor, and the two split the proceeds from the harvests. Sometimes these arrangements brought together different generations of the same family. For example, a colonist in Sinai tried to provide his landless son-in-law with a livelihood by allowing him to log the remaining tracts of land on the colonist's farm. Whether these arrangements spring out of a father's benevolent concern for the next generation or the labor requirements of a new cash crop, large households encourage the clearing of additional land.

Trends in the prices of agricultural commodities have also affected the rate at which colonists clear the remaining patches of rain forest. Initially, one might think of nibbling at islands of rain forest as an activity confined to poor households anxious to maintain a certain level of subsistence. The history of clearing forest remnants in the Upano—Palora region suggests otherwise. The pressure on the remaining islands of rain forest around Sinai did not parallel the increase in population in the village as one might expect in a process driven solely by demographic forces. The most concerted efforts to deforest occurred in the mid-1980s, when the price of narangilla rose sharply. The few landowners with woodlands near the road went into partnership with landless laborers, often younger brothers, to clear parcels of forest for narangilla cultivation. In Sinai a group of young, landless peasants proposed, without success, to use the village's common lands for narangilla cultivation. In Pablo Sexto the rising price of narangilla prompted a decision to divide up the forested common lands and give them to the landless sons and daughters of the original colonists. A number of the new landowners immediately began to clear land for narangilla cultivation.[26] The Shuar also clear more land when the price of narangilla rises. In 1991 narangilla prices hit new highs, and Shuar in the Upano valley cleared several small patches of primary forest to plant the fruit.[27] Population growth plays an indirect but important role in these instances of land clearing. It contributes to the large number of underemployed workers who move to clear land as soon as the markets made it profitable to do so.

The colonists' sensitivity to market signals has not always had a destructive effect on the forest. Impressed by the commercial value of the wood on their property, they have logged the forests, but they have also attempted to reforest their lands. The laurel (*C. alliodora*), good for making boards, sells particularly well in local markets, and a number of the colonists, having cut the laurels on their property, have attempted to replant them. Some have tried to plant laurels in open fields or as borders in fields while others have tried to plant them in both primary and secondary forests. The experiments in open fields have all failed, which is not surprising given that the laurel is not an aggressive, opportunistic species. It grows best in mature stands of trees. The colonists do not understand the difference between specialist species like laurels, which occupy niches in primary forests, and colonizing species such as balsa (*Ochroma lagopus*) and winchip (*Pollesta discolor*), which predominate in secondary forests. With their interest in laurels, the colonists have ignored the winchip, a less valuable but more common species in the increasing amounts of secondary forest around Sinai. There are no extension foresters in the region, so the colonists have nowhere to turn if they want help in reforesting the land. The few studies of forest ecology in Morona Santiago's rain forests indicate that the niches of commercially valuable species vary dramatically over small distances.[28] This pattern would complicate attempts to reforest the region with these species. Some smallholders want to reforest, but no one knows the silviculture of local trees, so no one knows how to reforest the land.

Growth coalitions, growing families, and rising commodity prices spurred land clearing on the Upano—Palora plain between 1970 and 1985. The coalitions mobilized local people and secured funds from national and international sources to finance the conversion of forests into fields. Both the colonists and the Shuar constructed coalitions, and both groups left legacies in the form of ties between lenders and small landowners that financed further deforestation in the region. The Shuar coalition, with its exacting organizational requirements and the uncertain commitment of its members to cattle ranching, has stimulated less land clearing than the colonist/CREA/BNF coalition.

The small size of our sample of individual landowners makes our conclusions about the individual factors that encourage the clearing of forest remnants somewhat tentative. Aside from a willingness to work with credit, increases in the density of the agricultural population, coupled with trends in commodity prices, appear to make the most important contribution to further deforestation. In the twenty years since they began farming in the region, many heads of households have watched their families grow from two to seven persons, and the growth in these households has spurred additional deforestation, especially among the Shuar. To provide for their children, elders had to subdivide their land when they passed it on to the next generation. The decline in acreage encouraged further deforestation, but the timing of the land clearing depended on trends in the prices of agricultural products. Population growth created a surplus of young laborers eager to clear land whenever price trends encouraged conversion. Chapter 7 focuses on the problems of these people, the second generation of settlers in the Upano—Palora region. It describes their search for a livelihood, their attempts to acquire virgin lands farther into the forest, and the implications of these efforts for future deforestation.

7
The Second Generation's Search for New Lands

■

LAND DEGRADATION IN AN OLD COLONIZATION ZONE

"By the time the road arrives in a place, the lands are tired."[1] With casual remarks like this one, colonists voice their concern about the growing problem of soil exhaustion in the Upano—Palora region. Pastures of gramalote (*Panicum purpurascens*), which used to mature in seven or eight months, now require nine to ten months to reach maximum height. Mature gramalote, which once grew chest high, now reaches a rancher's waist. In addition, the grass has changed color, going from a dark to a yellowish green.

The cattle find pickings slim in these pastures. Tied down by leashes, cows used to take an entire day to consume all of the grass within reach; now cows can eat all of the pasture within reach in the first three to four hours after the farmer ties them down for a feed. The growth curves of the cattle reflect the diminished quantity and quality of forage in Sinai. The mature cattle of the 1980s are smaller than their counterparts from the 1970s.[2]

Remote sensing analyses provide us with a means for assessing the validity of the small farmers' casual observations. Remote sensors can distinguish between pastures with healthy, mature vegetation and pastures with immature or stunted vegetation. The proportion of pastures in one or the other condition in a community provides a measure of the pressure of cattle on the land. Almost all of the fields in this region are pastures. When healthy pasture grasses reach maturity, they are shoulder high, and they have a lush, green color. Cattle move through fields in a concerted way because landowners leash their cattle in a line across the pasture and pull them into the mature grass once or twice a day. Between moves the cattle eat everything within reach, leaving behind a brown stubble. Because the stubble has very different reflectivity properties than green grass, the change in the appearance of the pastures registers in satellite images.

The stocking rate in pastures has some effect on the proportion of denuded pastures in a community. Landowners with more cattle per hectare of pasture may not wait as long as other landowners before they put their cattle into a pasture, so communities with high stocking rates would have a high proportion of denuded land. Soil exhaustion would also affect the extent of denuded land in a community because the rate of pasture regrowth depends on the level of nutrients in the soils. Grass returns more slowly in fields with depleted soils, so brown stubble predominates for a longer period of time, and satellite images record the persistence of the stubble. Regions with degraded soils should register more pasture with stubble than regions without degraded soils. Because the stocking rate affects the proportion of denuded land in a community, the satellite images do not give us a pure measure of land degradation but, because they do provide a partial measure, analyses of the images may suggest an emerging pattern of soil exhaustion in the region.

In some respects the pressure of cattle on pastures recalls the spatial coordinates of deforestation in the region. Only 27 percent of the pastures owned by large colonist landowners were in a denuded state compared with 44 percent of the pastures held by colonist smallholders. The large landowners not only clear a lower proportion of their

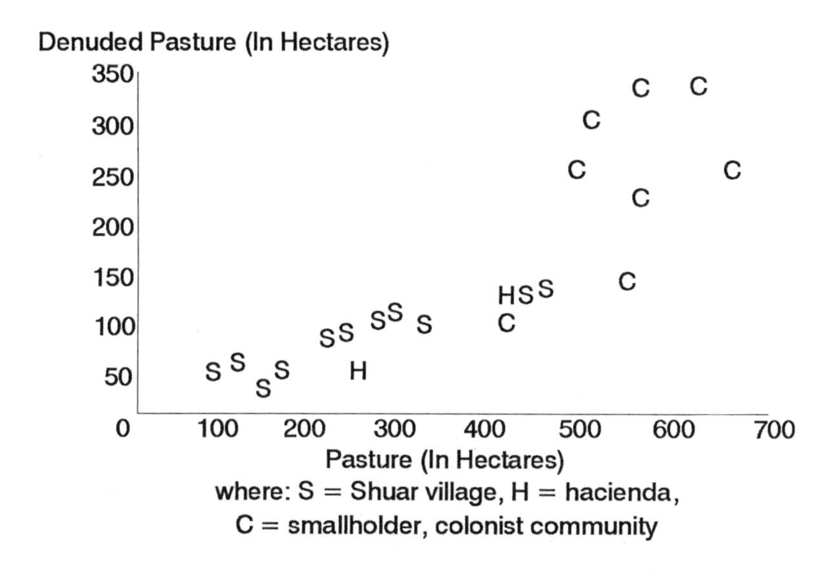

Denuded Pasture, Pasture, and Landowners' Ethnicity

lands, they also work their lands less intensively.[3] Other comparisons suggest that deforestation and pressure on pastures do not follow the same pattern. Deforestation varies significantly between Shuar and colonist communities, but the proportion of denuded pastures does not. While the colonists have cleared almost twice as much of their land as the Shuar, the denuded portion of their pastures is only slightly higher (41%) than the comparable figure (37%) for Shuar pastures. The scattergram in figure 7.1, which plots the amounts of pasture and denuded pasture in a community, suggests a curvilinear relationship between the two variables. Shuar communities with little pasture have a relatively high proportion of denuded land. Most probably, this pattern reflects frequently used garden plots near permanent village centers. The Shuar villages with the most pasture show only slight increases in the amount of denuded land, suggesting that they do not make heavy use of their pastures.[4] The largest amounts of denuded land are in the colonist communities.

Table 7.1. Denuded Pasture in the Upper Upano Valley, 1983

Panel A: Correlation Matrix

	(1)	(2)	(3)	(4)
(1) % denuded pasture				
(2) Distance from road	-.16			
(3) Date founded	-.49*	.21		
(4) Ethnicity	.05	-.44*	-.13	
Mean	38.9	5.1	1962	1.47
Standard deviation	12.2	6.7	7.3	.51

Panel B: Regression Analysis

% Denuded = -.33 = .062 Ethnic* + 2.19 Date + 1.26 Distance
(.023) (1.10) (1.94)

$r = .65$, r^2 (adj.) = .32, p + .02

N.B.: * = p<.05

As indicated in the correlation matrix in table 7.1, the places with the most denuded land tend to be the oldest communities in the region (r=-.49). They are all colonist communities near the provincial capital of Macas. After transforming the denuded pasture variable to correct for its skewed distribution,[5] we regressed the proportion of degraded land on three community characteristics.[6] The results, reported in table 7.1, indicate a pattern of degradation similar to the pattern of deforestation. The colonist smallholders, who clear the largest proportion of land, also work the cleared land the hardest. The survey data on the numbers of cattle per hectare in Sinai and Uunt Chiwias fit this pattern. The colonists have almost twice as many cattle per hectare as the Shuar (.60 to .33).

Smallholders talk about soil exhaustion, but so far it has not had dramatic effects on the cattle economy. Land prices have not declined, and established farmers in Sinai do not leave agriculture unless the BNF takes their land in foreclosure proceedings. The demand for extension services tells the story best. Gramalote, the most commonly used pasture grass, has low levels of nutrients, and the extension service provides more nutritious pasture grasses free of charge, but few smallholders have tried the new grasses. Instead, they call on extension workers for advice in curing sick cattle. Recurrent crises with sick cattle, which have an immediate impact on the colonists' incomes, consume all of their attention.[7] The long-term problems of soil degradation only get attention from the sons and daughters of Sinai's colonists. The diminishing returns to farming persuade many people in the younger generation to look for virgin lands farther into the forest rather than take over their fathers' farms.

A DIFFERENT GENERATION, A DIFFERENT SEARCH

When the landless sons of Sinai's founders despair at their chances of acquiring land, they remind themselves of Ebenecer. In 1978 two colonists from Sinai began looking for land farther east on the Upano—Palora plain. Having heard the Shuar talk about unclaimed tracts of land farther east, the colonists offered a Shuar acquaintance a considerable sum of money if he would help them locate an unclaimed tract of land to the east. The Shuar guide led the colonists to an expanse of unclaimed land between two *centros* forty kilometers east of Sinai. When the two men returned to Sinai, they quietly began to organize a group to occupy the lands. The group included their extended families and the new governor of the province, about thirty heads of households in all. Once the colonists had divided up the land into 100-hectare farms and cleared several tracts for pasture, they sold their farms in Sinai and moved to the new community. For his part the governor got the provincial council to appropriate funds for the construction of a 40-kilometer feeder road into Ebenecer from the Macas–Puyo highway (see figure 6.4). Five years after the initial settlement of the region, the new road reached Ebenecer. The original settlers, who sold off lots to later settlers, made large profits from the development of the new community.

Although the founders of Ebenecer worked with unusual speed and secrecy, their efforts resembled other settlement schemes in several ways. First, like at least three other groups of colonists in the Upano—Palora region, they employed Shuar as guides to help them locate pockets of unclaimed land. Second, almost all of Ebenecer's colonists came from other communities in Morona Santiago. In this respect the settlement of Ebenecer resembles the new wave of colonization occurring east of the Kutucu range around Taisha and Morona. Unlike early migrants to the mission settlements in western Morona Santiago who came from the Andean highlands, almost all of the migrants to the trans-Kutucu communities come from older settlements in Morona Santiago.

The change in the migrants' place of origin underlines a general change in patterns of migration in southern Ecuador. Peasants in the southern Andes no longer look east to escape poverty (ALOP-CESA 1984 2:207). Free lands no longer exist in the adjacent valleys on the eastern slope of the Andes, and close relatives do not have lands in the new colonization zones farther east, so highland peasants no longer learn about economic opportunities in newly settled regions. Cities rather than agricultural frontiers attracted the adventuresome youth of these villages during the 1980s (Rudel and Richards 1990). Less than 5,000 people migrated from the highlands to Morona Santiago between 1974 and 1982 (CONADE 1987:179).[8] Second generation colonists from the old colonization zones in Morona Santiago comprise the majority of the migrants to the new settlement areas.

THE NEW COLONISTS

Almost all of the old colonization zones eventually become sending regions for new waves of colonists who invade the forests of other, more remote regions. A wide variety of communities in the old colonization zones of Morona Santiago, ranging from places settled in 1920 to places settled during the 1960s, now send colonists to newly settled regions. Socioeconomic profiles of the Sinai and Uunt Chiwias residents who joined recent colonization schemes may explain how old colonization zones turn young men into migrants who push the boundaries of cleared land farther into forested regions.[9] The measures for the two dependent variables in these analyses are both dichotomous.

Desire for additional land? Yes or no.
Attempts to acquire additional land through colonization in the last ten years? Yes or no.

The measures for the three independent variables are outlined below.

Ethnicity: either Shuar or mestizo.
Household Size: the number of family members living with the respondent.
Income: an indirect measure derived from the income-producing assets on the respondent's farm; see below.

Devising a valid measure of household income that works with both Shuar and mestizo respondents proved difficult. Most Shuar households do not generate income on a regular weekly or monthly basis. To finance small purchases, they periodically sell plantains or coffee beans from their gardens. When an emergency requires a major expenditure, families raise money by selling assets. For example, they may sell a cow to pay for medical treatment when someone becomes seriously ill. In contrast most colonists generate a regular income through wage labor or the sale of farm products like narangilla or cattle. To measure income in a way which would yield valid comparisons between the two groups, we counted up the income-producing assets of a household, including wage labor, and estimated the income which these assets would produce over the course of a year. The heads of households answered questions about their sources of income and the extent of their assets. How many cattle did they own? How much acreage did they have in coffee and narangilla? How much did they earn as wage laborers in the last year? Informal conversations with knowledgeable residents produced consensus estimates of the income which, for example, a herd of 30 head of cattle or a hectare of narangilla would produce during a year. Finally, the household heads estimated the proportion of their income derived from their most important on-farm resource. From these data we could calculate household income for the year. These estimates represent average incomes over the course of several years.[10]

The regression analyses in table 7.2 suggest some of the forces that lie behind the quest for more land.[11] Here too, the pressure of numbers makes itself felt. Large households not only clear a larger proportion of their land, they also try to acquire additional land more frequently than do small households. Fathers participated in colonization schemes because they wanted to obtain land for their second and third sons. Income also made a difference. Affluent households joined these ventures more frequently than poor households because failures

Table 7.2. Attempts to Acquire New Lands

Panel A: Attributes of Smallholders in Two Communities

Variables	Means: Colonists	Means: Shuar
Household income (US$)	2713	1308
Household size	6.8	5.9
Acquired land through colonization (% yes)	83%	28%
New lands, desire (% yes)	7%	49%
New lands, attempts (% yes)	53%	9%

Panel B: Correlation Matrix

Variables	(1)	(2)	(3)	(4)	(5)
(1) New lands, desire					
(2) New lands, attempts	.558				
(3) Household size	.609	.389			
(4) Ethnicity	-.082	-.481	-.028		
(5) Income	.195	.467	.240	-.538	
Mean	1.47	1.62	6.37	1.51	1998
Standard deviation	.50	.49	3.05	.50	1118

Panel C: Logistic Regression Equations

Equation #	Eq. (1) Desire	Eq. (2) Desire, Shuar	Eq. (3) Desire, Colon.	Eq. (4) Attemps	Eq. (5) Attempts, Shuar	Eq.(6) Attempts, Colon.
Variable						
Household size	.325*	.282*	.365*	.172*	.252	.361*
	(.088)	(.110)	(.134)	(.067)	(.976)	(.137)
Income	.471*	.021	.013	.960*	.310	.360*
	(.313)	(.026)	(.019)	(.431)	(.300)	(.230)
Ethnicity	.571			.625*		
	(.412)			(.340)		

N.B.: * indicates that the regression coefficient is at least one and half times its standard error.

did not have as severe financial repercussions for them as they did for poorer peasants. The organizers made it easy for the wealthy to participate by exempting them from the punishing work of trailblazing if they made large enough financial contributions to the coalitions.

The Shuar and the colonists differed in the frequency with which they joined schemes to acquire more land. While similar proportions of colonist and Shuar households, 57 to 49 percent, wanted more land, they differed sharply in the proportion of households, 53 to 9 percent, which invested their own resources in efforts to acquire new lands (see table 7.2). To some extent the greater participation of colonist households in these schemes reflects their greater wealth. Older colonists, with modest accumulations of wealth, made financial contributions that supported the young while they established claims to forested land. With incomes averaging only 48 percent of the colonists' incomes, most Shuar could not afford to support their young in pioneering ventures.[12]

Previous experiences made the colonists eager and the Shuar reluctant to participate in new land settlement schemes. Most colonists had memories of a successful move from the Andean highlands to their present home in the Oriente. Having succeeded once in the acquisition of new lands, they see real possibilities for acquiring additional lands in new colonization zones. In contrast most Shuar in western Morona Santiago acquired land through reorganization rather than through a long distance move to a new site. In the older *centros* many of the younger landholders inherited their land. Without a positive personal experience with colonization efforts, the Shuar in Uunt Chiwias usually said that they wanted more land, but they seemed reluctant to look for additional lands, arguing that Shuar and Achuar to the east had already claimed all of the arable land. In contrast the colonists often launched individual and group efforts to claim land with only a cursory search for competing claims.

The Shuar's reluctance to pioneer may also reflect the continued availability of land for cash cropping in Uunt Chiwais. Second and third generation Shuar typically acquired at least 15 hectares of land through inheritance, and a landowner could earn a small income growing cash crops on a tract of land this size. Land scarcity figured prominently in the decisions by many Shuar to move from the more densely settled *centros* near Mendez to the unclaimed tracts of forest south of Coca (Descola 1981:640).[13] Acculturation may also play a role in the Shuar's reluctance to move. Despite the group's history of dispersed settlement, many young Shuar will only establish a new home in the forest as part of a larger group headed by older Shuar. Young, educated Shuar seem particularly reluctant to establish new homes in the rain forest. The more schooling the young Shuar have, the less they know about the flora and fauna of the rain forest, which in turn makes them more reluctant to establish new homes in an unoccupied tract of forest.[14]

The younger generation of colonists sought new lands in a variety of locations. In some instances they staked out claims close to home along the edge of the Macas–Puyo corridor of cleared land. In other instances they claimed large blocks of land far from the sending communities. Young men from Sinai did both during the 1980s; they pushed the fringes of the forest westward toward the Andes, and they claimed blocks of land far to the east of the settlement corridor. Each type of agricultural expansion posed special problems for colonists.

WIDENING THE CORRIDOR OF CLEARED LAND

Establishing the Claims

When the first settlers in Sinai had cleared away enough forest to see the western horizon, an awesome sight unfolded before them. A range of mountains stretched from north to south as far as the eye could see, some 12,000 to 14,000 feet above the jungle floor. Three snow capped peaks topped the range.[15] As the settlers' eyes traversed the forty kilometers between Sinai and the mountains, green forest gave way to blue foothills and then to black volcanic rock at the crest of the ridge. For several years the colonists did little more than marvel at the view but, as unclaimed land close to Sinai became scarce, they began to wonder how much arable land lay between themselves and the mountains.

The colonists began to push the margins of settlement toward the mountains shortly thereafter. In 1975 and 1976 CREA persuaded groups of colonists from Azuay, organized into La Quinta and La Sexta cooperatives, to settle to the west and north of Sinai. Meanwhile, colonists in Sinai expanded to the northwest across the

Ambusha River (see figure 6.4). Between 1972 and 1976 the colonists established more than forty new farms on the far side of the Ambusha. Farther north, two groups of CREA assisted colonists, La Septima and La Octava cooperatives, established settlements west of Pablo Sexto. Not satisfied with the allotments of land granted them within the cooperatives, a group of Pablo Sexto colonists, along with the governor, IERAC officials, and influential families from Macas, formed Alianza Agricola.[16] This group of seventy persons claimed 70 hectares apiece west of La Septima community. Two smaller groups of colonists, numbering about thirty persons each, claimed lands northwest of La Quinta cooperative during the late 1970s and early 1980s.

In the mid-1970s the National Park Service complicated the colonists' plans for westward expansion by creating a park, Parque Nacional Sangay, around the three snow-capped peaks and their foothills. The natural beauty of the place, the presence of endangered species, and the absence of human settlements made the Sangay region an appropriate site for a park.[17] Budgetary surpluses in the national capital and news of the colonists' westward advance toward the mountains spurred the planners into action.[18]

When news of the park's creation reached Sinai, the colonists worried that the government would seize their lands because they fell inside the park's boundaries. In 1981 after several years of lobbying by the colonists, park officials drew new boundaries, closer to the mountains, which excluded the colonists' claims and reduced the size of the park. In 1982 the park service hired a colonist to cut the park's eastern boundary line. The park service promised to pay the young workers who cut the boundary with grants of land immediately to the east of the boundary. The colonist contractor began drawing an irregular boundary line with deviations to allow his friends to claim choice parcels of land. When the park's chief discovered the deviations, he fired the contractor, retained the workers, and cut a boundary line due north from the narrows on the Upano River. Because politically influential members of Alianza Agricola claimed land along one section of the park boundary, some of the workers on the boundary cutting crew never received any land for their labor.[19] Despite rugged terrain featuring steep hills and deep ravines, all of the land between the colonists' villages and the park, a distance of approximately twenty kilometers, had been claimed by 1984.

The Limits of Expansion

Although the colonists claimed the uplands quickly, they cleared it slowly. Living conditions in these isolated places slowed the pace of development. Colonists would typically work on their claims for four or five days in a row until they ran out of food. During rainy periods hungry colonists returning to Sinai at the end of the week would arrive at the banks of the Ambusha to find their path blocked by the swollen river. Rather than wait for the waters to recede, the colonists would drop a tree across the river and use it as a bridge to walk or crawl across the raging torrent. These makeshift bridges would often last only a few hours before the surging currents would sweep them away.[20] The power of the flood waters underscored the hazards of the river crossing and reduced the colonists' enthusiasm for working lands on the far side of the river.

In addition to the hazards of travel into the new claims, the colonists had to endure isolation from family and friends in order to work the land. The children remained in the village to attend school, and the mothers stayed with them, so the men lived alone on their claims during the week. Some colonists endured longer periods of separation. Many of the women and children of families in La Quinta cooperative continued to live in the Sierra for years after the colony's creation, so the men lived alone on their claims. The members of La Sexta cooperative conveyed the dispiriting effects of isolation in the name they choose for their new community. They called it "Siberia Amazonico."

Faced with the difficulties of establishing a farm in a remote place, many colonists lost heart and abandoned their claims. In an area bordering the national park, approximately twenty colonists claimed land in the early 1980s, but only four or five of them worked the land, and even then they cleared little land. In one typical case a young colonist cleared 10 hectares out of a 90-hectare claim. Only one colonist has built a house on lands adjacent to the park. The amount of cleared land increased with proximity to the village of Sinai. A remote sensing analysis of the 1983 satellite image compares land use near the village of Sinai with land use on the periphery, both to the west, across the Ambusha River, and to the north, in La Sexta community. It reveals substantial differences in the extent of deforestation. While the

Table 7.3. Sinai Farms by Distance from the Road

Variables	Farms Around Town Center (close to road)	Farms Across the Ambusha (far from road)
Landholds (ave. in hectares)	50	30
% in pasture	73.7%	67.2%
# of cattle	28	7
cattle/pasture	.80	.38
# of cases	16	10

SOURCE: Calculations based on survey results in T. Yaccino 1988:54.

colonists had cleared 55 percent of the land around Sinai, they had deforested only 29 percent of the land beyond the Ambusha and 32 percent of the land in La Sexta.[21]

A 1987 survey of colonists with farms close to and far from the road provides another means of assessing the importance of location in land clearing. The farmers in remote locations, across the Ambusha River, began their farms several years after the colonists in central locations, near the village; the Ambusha farmers received only 30 hectares of land as opposed to 50 hectares, and their farms are located one- to three-hours' walk from the road rather than on the road. Because the Ambusha farmers have smaller tracts of land than the original colonists, they might be expected to clear a higher proportion of their land. The data in table 7.3 do not conform to this expectation: the colonists close to the town center cleared a greater proportion of their land than the colonists across the Ambusha.

While the differences in the extent of deforestation were not dramatic, the differences in the size of herds were substantial. The original colonists around Sinai had four times as many cattle as the colonists across the Ambusha. The stocking rate (cattle/pasture) on the farms close to the road was twice the rate for the farms far from the road. Because the pressure of cattle on existing pasture largely determines a farmer's plans for future land clearing, the low stocking rates on the Ambusha farms implied low rates of deforestation on these farms. The low stocking rates also suggest that the Ambusha farmers did not have the capital to purchase cattle. By making it difficult to purchase more cattle, the farmers' poverty prevented additional agricultural expansion in the Ambusha region.

A 1986 survey of occupations sheds some light on the poverty of farmers far from the road. Nearly 40 percent of the original colonists have a second occupation, usually as middlemen, which provides them with another source of income while only 3 percent of the colonists who arrived later and settled beyond the Ambusha have a secondary business.[22] The location of the original colonists' farms and homes along the road in Sinai makes it possible for them to work as middlemen. The late-arriving colonists have homes in the town center, but their farms are located two- to three-hours' walk from the village center. To work the land, the late arrivals must live on their farms during the week; on weekends they return to the village to see their families.

Because their agricultural work takes them away from the town center and its road, these farmers cannot develop secondary businesses as middlemen. In contrast the original colonists with farms close to the town center spend every day of the week in town so they can run small businesses there. The original colonists buy, sell, and transport goods; they purchase narangilla and wood from other colonists, run one of the seven stores in the parish, or transport goods and people to Macas in their trucks. Among their steadiest customers are the peripheral farmers across the Ambusha and the colonists in the outlying communities founded by La Quinta and La Sexta cooperatives. This system of marketing requires that peripheral farmers endure price gouging by an additional set of middlemen in order to work on their farms. Under these circumstances peripheral farmers cannot amass the capital necessary to purchase additional cattle, so they have no reason to clear more land for pasture.

All but the most affluent colonists must work periodically as wage laborers in order to support their families. During the first two years after settlement 30 of the 42 household heads in La Octava cooperative spent large amounts of time working on a regular basis as wage laborers in neighboring communities.[23] Five years after the initial settlement of La Quinta, more than half of the landowners would leave for up to six months at a time to live and work in their communities of origin in the Sierra (Salazar 1986:72, 204). The extensive off-farm work made it difficult for colonists to convert extensive tracts of forest to pasture, so they could not maintain large herds of cattle on their farms. Given the small size of the herds, the farmers could not sell enough cattle to support their families, so they had to continue working elsewhere as wage laborers.

By the early 1980s land consolidation had begun throughout the peripheral regions to the west of the Macas–Puyo highway.[24] Colonists became discouraged and sold out to their neighbors. Because farmers with larger units almost always clear a smaller proportion of their land than farmers with smaller tracts of land, the consolidation of farms implied a reduction in the eventual extent of deforestation in the region. The departure of colonists on outlying farms suggests that by the early 1980s the corridor of cleared land along the Macas–Puyo highway had reached the economic limits of its expansion. Corridors of cleared land do not continue to widen indefinitely. Problems of access become more difficult; people begin to look elsewhere for their livelihoods, and the land remains forested.

Commodity Prices, Public Works, and Expansion on the Periphery

Price trends during the mid-1980s promised renewed agricultural expansion along the edges of the Macas–Puyo corridor. In 1985 and 1986 the price of narangilla reached all time highs. The price increase stimulated interest in the uplands west of the colonies for several reasons. First, narangilla requires well-rested soils for optimal yields, and the western edges of the colonist communities had ample supplies of virgin land. Second, narangilla yields larger volumes of fruit in subtropical rather than tropical climates, so cultivation of narangilla in the colder climate of the foothills west of the colonies promised to yield extraordinary quantities of the fruit.

Several IERAC officials understood this calculus and invested in lands west of Sinai during the mid-1980s. Young colonists in Sinai also showed renewed interest in the uplands and purchased lands near the park. Several older colonists cleared land and planted narangilla in the foothills at the same time that other colonists cleared land for narangilla cultivation at lower elevations near the road. No one made a concerted effort to improve transportation into the foothills while commodity prices were high, and the economic rationale for doing so faded quickly. By 1989 declines in the price of narangilla, coupled with the collapse of the bridge over the Upano, had made it impossible to market narangilla from the upland plots and make a profit. Several landowners tried to sell their lands in response to the adverse trends, but they could not find buyers. The market for land in these places had collapsed.

To the west of Pablo Sexto a similar story unfolded. During the oil boom from 1973 to 1977 CREA expanded the region's infrastructure. The agency constructed footbridges over the main rivers and laid down *empalisadas* on most of the main trails. In the late 1970s CREA's financial condition began to deteriorate, and it tried to transfer the costs of trail and bridge maintenance to the provincial government, but the province refused to accept the added responsibility. CREA could no longer afford to maintain the trails and bridges, and they began to deteriorate. By the mid-1980s the trails to La Septima and La Octava communities from Pablo Sexto had deteriorated so badly that parents in the two outlying communities no longer allowed their children to walk to and from school in Pablo Sexto. To keep their children in school, the families moved to Pablo Sexto.

During the same period land consolidation began around the two villages as discouraged colonists sold out to more persistent neighbors. Alianza Agricola, the group with influential members who had claimed land west of La Septima, applied to the provincial government for funds to construct an *empalisada* into their lands. The province did not provide the funds, so access remained difficult, and people cleared little land. Rising narangilla prices caused some deforestation, but all of it occurred in islands of forest close to the road in Pablo Sexto.

In these instances rising commodity prices did not induce road construction and additional deforestation. A crisis in the federal government's fiscal condition, coupled with local political considerations, made funds for feeder roads difficult to obtain when narangilla prices rose in the mid-1980s. Problems with Ecuador's external debt forced the country's political leaders to curtail road construction projects, including the Macas–Puyo penetration road. Because the penetration road would link Morona Santiago with an extensive road network in the northern Oriente and Sierra, the provincial council felt a political obligation to continue the road's construction, so they decided to replace the lost federal funds with the council's funds. The diversion of council funds to the penetration road reduced the availability of funds for feeder road construction and stopped the westward expansion of settlement on the Upano—Palora plain.

These circumstances raise several questions about growth coalitions. Are the coalition members influential enough to get the government to commit the funds to build a feeder road? If their initial efforts to obtain funds for a road do not succeed, does a coalition have the cohesion necessary to sustain a prolonged lobbying effort? When coalitions cannot obtain funds for feeder road construction, they usually contribute only marginally to rain forest destruction.

CLAIMING BLOCKS OF RAIN FOREST

When peasants organize to claim and clear blocks of rain forest, their success hinges on the availability of large tracts of rain forest. A review of recent attempts by groups to acquire and clear large areas underscores the importance of a supply of unclaimed land for the success of these group ventures.

Via Curaray. In the early 1980s large numbers of Shuar from the old, densely populated *centros* near Mendez began selling their land and moving north to the province of Napo. There they established twelve villages in the path of a road being built by a French company, Elf Aquitaine, into a recently discovered oil field along the Curaray River (see figure 3.1). Unlike the mestizo colonists in the same region the Shuar founded their villages about five kilometers into the forest on either side of the advancing road. The younger Shuar preferred to be closer to the road where they could grow coffee for sale in urban markets. In some instances the Shuar settled on lands reserved for another indigenous group, the Huaorani, and they did so without the support of the federation.[25]

Nangaritza Valley. Young Shuar from the densely settled *centros* near Gualaquiza have established *centros* in the upper reaches of the Nangaritza valley near the disputed border with Peru. Colonists from Loja have also established settlements in this area.[26] To secure Ecuadorian sovereignty over the region, the central government began an integrated rural development program, which includes the construction of a penetration road up the valley toward the lands that the Shuar and the colonists have claimed for their new communities.

Centro Sangay. In the early 1980s a group of Shuar from the densely populated village of Wapu on the Upano River moved upstream into the national park and established a new village, Sangay, inside the park's boundaries. The Shuar found the hunting to their liking, and cleared about 20 hectares of land for gardens. Eight months after the Shuar founded the village, the park rangers discovered it and evicted them.[27]

Nuevos Horizontes. Colonists formed this group in 1985 when several public employees, former colonists themselves, made a reconnaissance flight over an area with broken terrain in the foothills of the Kutucu range to the south of Ebenecer. The organizers argued that their own extensive experience with colonization would improve the project's chances of success. Fifteen years earlier, they had participated in the founding of Sinai and other Upano—Palora communities. Now they had purely speculative interests in the land. As one disenchanted colonist observed, "the organizers will never put on boots." Later, a Shuar guide led an exploration party into the region and a group of landless peasants, some from Sinai, cut boundary lines. The provincial council provided foodstuffs for the workers, and the older, wealthier members of the group contributed funds that the workers used for supplies.

A large number of peasants made trips into the region to look at the lands. Some of them got lost and gave up. Others reached the lands, but many of them decided not to join the group. To curb the "tourism," the group's organizers began requiring a US$15 membership fee before someone could inspect the land. Although the young members of the group made a series of trips into the land, the project eventually collapsed.

Three factors explain the collapse. The nearest Shuar village had already claimed an extensive portion of the land, raising the prospect of a lengthy political battle over titles to the land. The limited amounts of flat land in the zone discouraged potential colonists. They assumed that the group's leaders would claim the flat, arable lands for themselves and leave the rough terrain for the peasants. Finally, in promoting the colonization scheme, the leaders of the council promised funds for the construction of an airport on the land. When the council did not provide the funds, many participants became discouraged, and the group disintegrated.

Bobonaza. The brother of Sinai's mayor organized this group in 1988 after talking to some Shuar who wanted to sell land about three-hour's walk east of the Pastaza ferry crossing. The completion of the Macas–Puyo road and the inauguration of ferry service across the Pastaza triggered a sharp increase in land prices, and some Shuar wanted to take advantage of the price increases by selling their portion of their *centros'* lands and moving elsewhere. Approximately 200 persons, including merchants from Macas and Cuenca as well as colonists from Sinai, formed a group, cut trails, and built houses on the land. Other *centro* residents, backed by the federation, objected to the land sale, and a dispute began. It intensified until the rival claimants began shooting at one another. The opponents of the land sale prevailed in the conflict among the Shuar, and the colonists abandoned their claims to the land after the Shuar owners withdrew their offer to sell the land. The members of the colonist group lost around US$200–300 per person in this venture.

These histories demonstrate that the larger group ventures also depend for their success on the continued construction of infrastructure in a region. When the building slows down as it did in the Upano—Palora region during the 1980s, large blocks of unclaimed, accessible lands quickly disappear, and groups encounter opposition from existing owners when they try to occupy a large tract of forest. Opposition from competing claimants accounts for the failure of the Centro Sangay, Nuevo Horizontes, and Bobonaza projects.

The Via Curaray and Nangaritza valley projects solved the access problem by "piggybacking" on an institution's road building plans and only in these instances did a group venture produce additional deforestation. Without the continued construction of new roads, groups quickly occupy all of the accessible rain forest, and their presence frustrates attempts by later groups to occupy lands in this area. A continuous supply of new roads insures a continuous supply of unclaimed lands, which contributes to the success of group ventures and a continuation of rapid deforestation in a region.

In some exceptional instances individual colonists, experienced in colonization, succeeded in acquiring additional lands through a free rider strategy. Men from two large families in Sinai established a claim to lands near Morona, the terminus of the recently completed Mendez–Morona highway. Again, competing claims, established in this instance by colonists from Sucua, made the eventual outcome of the pioneers' work uncertain. In another instance a poor Sinai family lost their land to the bank and moved to Villano, the anticipated terminus of a road under construction north of the Pastaza River. Following the example set by the more prosperous colonists in Sinai, the family hoped to claim land in the path of the new road and profit from the expected appreciation in land values when the government completed the road.

In these examples the colonists, like the Curaray and Nangaritza Shuar, did not join coalitions that included people who could build roads. Instead, like players in a game, they anticipated the actions of road builders and claimed lands whose access to the outside world would improve with the construction of a road. Whether or not someone seeks additional lands individually or in a group, the provision of infrastructure by more powerful entities remained a requirement for success in acquiring and clearing large tracts of rain forest during the austere economic times of the late 1980s. During periods of austerity, when growth coalitions have difficulty acquiring funds, road building by lead institutions becomes an even more important source of support for land clearing.

THE PROBLEM OF THE SECOND GENERATION

The declining possibilities for agriculture on the fringes of the Macas–Puyo corridor, coupled with the dismal record of recent growth coalitions in acquiring land, underscore the difficulties the sons and daughters of the original colonists face in securing a livelihood in the Upano—Palora region. While recent migrants from arid highland valleys continue to marvel at the lush tropical growth, the young people who have grown up in the region see a somber future for themselves in agriculture. Because almost all of them come from families with two or more sons, they will inherit subdivided lands;

they will have to generate an income from 25 rather than the 50 hectares their fathers had. In addition the inherited lands are often "tired," unable to support a narangilla grove or grow abundant pasture for cattle.

The young also bring a different set of values to farming than did their fathers. While the first generation of colonists celebrated prowess with an ax, members of the second generation look at woodsmen with indifference. More education and exposure to urban attractions has given the second generation a different set of values. In the words of one middle-aged colonist, the young "go to school, and they learn to be lazy, so they do not want to get muddied up walking the trails or tending cattle."

The values of the second generation have changed, but urban economies have not created suitable jobs for them. In numerous instances the sons of Sinai's original colonists have turned down their aging fathers' offers of a partnership on an *a medias* basis in the cattle ranching enterprise, but they have not committed themselves to an alternative occupation. In some instances they have settled for part-time employment. One son of a prominent colonist sold his land across the Ambusha and moved to the provincial capital where he maintains a volleyball court and sells soft drinks when young men gather to play in the late afternoon. On occasion he cuts wood on a contractual basis in the remaining rain forest along the penetration road north of the city. In another family the son wants to drive a bus, and he now has his license, but the family cannot afford to buy a bus, so he sits at home with little to do.

Slightly older men, with families to feed cannot afford idle time, so they engage in a continual, sometimes frantic search for income. Their search leads to involvement in a bewildering variety of enterprises. The occupational history of Miguel D., the son of a Sinai colonist, illustrates the pattern.

Miguel D.: An Occupational History

Born into a family with some capital and several plots of land in a highland community, Miguel moved to the Oriente as a teenager in the early 1970s. When his father and older brother acquired land in the Upano—Palora region during the late 1960s, he was too young to obtain lands by joining a cooperative. Miguel graduated from an agricultural high school in the mid-1970s and began working in a menial position at an agricultural experiment station outside of Quito. After several years the low salary and poor prospects for advancement discouraged him, and he returned to the Oriente to care for the cattle on his father's 50-hectare farm.

Several years later, now married, Miguel became involved in a scheme to reopen a gold mine in the Sierra, and he spent long periods of time away from his wife and young sons working in the mine. After the mining venture failed, he returned to the Oriente and joined the Nuevos Horizontes colonization group. A week-long trip into the claimed area convinced him that the terrain was too rugged to farm, so he withdrew from the group.

During the narangilla boom he led an effort by young, landless men in Sinai to get permission to grow narangilla in the community's 30-hectare forest reserve. When this initiative failed, he and his older brother decided to grow narangilla themselves. His brother supplied the land, and he provided the labor. With the declining prices for narangilla they made little money from this venture, and Miguel turned to other pursuits.

In 1988 he joined other young men from Sinai in the abortive attempt to acquire lands near the Bobonaza River. Miguel's father, Don Alfonso, has built up a herd of almost fifty head of cattle. Failing of eyesight, Don Alfonso would like Miguel to take over the care of his cattle on an *a medias* basis, but Miguel refuses to do it, arguing that he cannot feed his family on the proceeds from the farm.

More recently, Miguel and an artisan in Sinai began making parrot and parakeet figurines out of balsa wood and selling them to tourist shops in Quito. Miguel is now 34 years old, and he has four children to feed and cloth. In late 1989 he was seriously considering selling his motorcycle and using the proceeds to travel to the United States to look for work.

The rapid succession of jobs and enterprises in Miguel's life exemplifies the elasticity required of so many marginalized peoples whose livelihood depends on trends and decisions emanating from metropolitan centers (Geertz 1963:103). Faced with dramatic shifts in the array of external forces, the younger colonists in the Upano—Palora region have few established routines or institutions to preserve, so they respond by moving away. Miguel is not alone in thinking about emigration to the United States. Two young men from the area have already gone, and many others talk of going. These discussions suggest the extreme labor mobility that prevails among the second generation of colonists. The prospect of moving does not seem unusual to them. Their parents made an economically rewarding move to a rain-forest frontier in the Oriente, and they come from a region in the Sierra where for decades men migrated to the coast to work on plantations.

In sum the sons and daughters of the original colonists, more than their peers in nearby Shuar communities, regard migration as an important means of responding to economic opportunity. The original colonists moved to the rain forest to start a farm, and many second generation colonists have tried to make a similar move in more difficult economic circumstances. If the returns to colonization decline even further, these people will move elsewhere, perhaps to urban areas in other parts of Ecuador. Recent changes in interprovincial migration support this expectation. From the 1962–74 to the 1974–82 intercensal period, net migration into Morona Santiago declined substantially. While the proportion of in-migrants in Morona Santiago's population declined slightly, from 34.9 percent of its population in 1962 to 29.8 percent of its population in 1982, the number of emigrants from Morona Santiago as a proportion of the province's population increased from 5.4 percent in 1962 to 12.2 percent in 1982 (CONADE 1987:158). Many of the young migrants leaving the old colonization zones, the Cuyes Valley as well as the Upano—Palora region, moved to urban areas in the Sierra like Cuenca or Quito.[28]

The establishment of Ebenecer during the late 1970s provides a sterling example of how successful growth coalitions contribute to the destruction of large areas of rain forest over a short period of time. In a pattern consistent with colonization efforts in other parts of the tropical world, the people who joined these second generation growth coalitions came in disproportionate numbers from the largest and the wealthiest families in the Upano—Palora communities. The size of these families created the need for more land, and the wealth provided the means for finding it.

In recent years most of these colonization efforts have failed because the pace of road building has declined. Without additional road building to open up new areas, groups quickly occupied all of the accessible land, and later group ventures rarely ever succeeded because, to do so, they had to wrest control of the land from the first groups. In many instances individual colonists obtained land along the fringes of the Macas–Puyo corridor, but continuing difficulties with access prevented the deforestation of these lands. Taken together, the recent experience of these colonization groups contradicts the claim that tropical deforestation is an inexorable process. At best it occurs in fits and starts.

The same dispositions and circumstances that make many colonists so ready to join with others in the search for new land also predispose them to move out of tropical regions altogether if opportunities warrant. The high degree of labor mobility presents policymakers intent on reducing rates of deforestation with both a threat and an opportunity. Peasants will continue to flock to the corridors surrounding penetration roads if companies and countries continue to build roads, but, if the road building ceases, so will a lot of the deforestation because people will look outside of the rain-forest regions for economic sustenance. The following chapter addresses these policy issues in more detail.

8

Tropical Deforestation: An Assessment with Policy Implications

This book has argued that tropical deforestation typically takes the three forms outlined in table 8.1. Two types occur in places with large blocks of rain forest. One of them features a lead institution that opens a region up for development, usually by building a penetration road into the forest. Free riding groups and individuals take advantage of the increased access, stake out claims to land, and begin to clear it. In the other pattern peasants, investors, and government officials pool their resources in coalitions that develop rain-forest regions for agricultural production. The internal structure of the coalitions varies considerably; some coalitions, especially those based on kinship, involve a union of equals; other groups have a leader around whom its members coalesce. In most coalitions the members divide up the labor in uneven, but understandable ways. In many coalitions the leader provides the infrastructure while the peasants, with help from

Table 8.1. Varieties of Tropical Deforestation

Size of Forest	Process	Precipitating Factors [a]
Large: blocks of forest	Lead institutions; free riding peasants	Natural resource extraction; demographic pressures; rising agricultural prices and proletarianization
Large: blocks of forest	Growth coalitions	Funds for infrastructure; demographic pressures; rising agricultural prices and proletarianization
Small: islands of forest	Encroachment	Demographic pressures; rising agricultural prices; proletarianization

[a] The precipitating factors are listed in their order of importance in initiating each of the three processes of deforestation. The importance of these factors will vary from place to place. These judgments are based on reports of deforestation in Africa, Asia, and Latin America.

investors, occupy and clear the land. A series of these coalitions changed the landscape in Morona Santiago between 1920 and 1990.

In the third pattern, which characterizes places with small, remnant patches of rain forest, cultivators nibble at the edges of the forest. Working alone or with family members, farmers expand their fields at the expense of the forests. The present chapter summarizes the three patterns, addresses questions about their prevalence in rain-forest regions throughout the world, and spells out their implications for the design of policies to reduce rates of deforestation.

THE CASE STUDY

The history of development and deforestation in Morona Santiago provides multiple examples of growth coalitions that developed and deforested places. When priests and colonists joined forces to build agricultural communities around missions during the first half of the twentieth century, they cleared large areas in the valleys at the base of the Andes. When a regional development agency began to build roads in Morona Santiago during the 1960s, knowledgeable individuals formed coalitions to occupy and exploit the newly accessible lands. Public officials, colonists, and investors from the Sierra joined forces to open up the Upano—Palora region for cattle ranching during the late 1960s. Twenty years later some of the same people cooperated in schemes to develop the more remote parts of the Upano—Palora plain. The Salesians and Shuar leaders formed counter-coalitions that secured global titles to land and channeled funds from European foundations to those Shuar intent on converting forests into pastures for cattle.

The coalitions' contributions to land clearing varied with their composition and the context within which they worked. Shuar ambivalence about deforestation made their coalitions less effective than colonist coalitions in clearing land. Kin-based coalitions of Cuyes valley colonists cleared land slowly, but they persisted in clearing land even after a development agency withdrew its support. Coalitions with an agency's support, like the Upano—Palora groups, cleared land more rapidly. Changes in the macroeconomic context affected the coalitions' ability to raise capital and clear land. With the abundance of capital that accompanied the oil boom in the 1970s, coalitions could secure funds for road building, so they succeeded in opening up areas for settlement, and deforestation increased. A dearth of capital, as occurred in the late 1980s, slowed the road building and deforestation. Even with rising prices for agricultural commodities, colonists cleared little land.

The rapid land clearing of the 1970s fragmented the forest on the Upano—Palora plain and changed the dynamics of deforestation. Individual factors became more visible as determinants of deforestation. Some smallholders secured credit from the banks more easily, and they cleared larger proportions of their land. Some families had more children than other families, and the larger families cleared a greater proportion of their land.

These differences in household size also had an impact on deforestation in distant places. Large households joined coalitions searching for new lands more frequently than did small households. Although the pressures of providing for large families persuaded some fathers to look for new lands, agricultural expansion did not involve a desperate attempt to maintain a certain level of subsistence. The households which searched for new lands, in addition to being larger in size, had higher incomes than other households. The heads of these large, relatively affluent households talked about the acquisition of new lands as an investment in the younger generation rather than as an attempt to maintain their standard of living. The entrepreneurial inclinations of these peasants implies high rates of geographical mobility. Having moved once with some success, they will move again. The sons and daughters of the original colonists in Sinai are just as likely to move to an urban area as they are to move farther into the rain forest.

While the history of colonization in the Upano—Palora zone provides a clear example of the two-stage sequence in land clearing, the increased salience of the urban alternative for the second generation suggests the historical limits of this generalization. By changing the array of choices available to the younger generation, urbanization has affected decisions about clearing additional land. A comparison between Cuyes River colonization during the 1960s and Upano—Palora colonization during the 1980s illustrates this effect. In both places road construction proceeded slowly, but the Upano—Palora colonists seemed more sensitive in the 1980s to changes in the pace of construction than the Cuyes valley colonists were during the 1960s. The Upano—Palora colonists scrapped their plans to clear additional land as soon as road construction and trail maintenance ceased in the early 1980s. During the 1960s the construction of the Gima–Cuyes penetration road started and stopped, but the colonization of the Cuyes valley continued.

The different reactions of the two groups of colonists to changes in road building plans may reflect the increased importance of urban labor markets during the later period. During the 1960s the much smaller urban economies did not provide peasants with an alternative set of opportunities, so they committed themselves to colonization and deforestation. The presence of a larger urban economy and the Upano—Palora peasants' greater familiarity with it may have made them more sensitive to marginal changes in the economic opportunities afforded by colonization in the 1980s. When the growing inaccessibility of the remaining unclaimed lands diminished the attractions of colonization, the Upano—Palora colonists immediately began to explore urban economic opportunities. If unclaimed lands become even more inaccessible and urban expansion continues, fewer examples of the first-stage group ventures would occur, and the two-stage sequence would become a less accurate description of how deforestation occurs.

HOW WIDESPREAD IS THIS PATTERN?

The case study provides numerous examples of how growth coalitions stimulate tropical deforestation, but questions remain about the number of rain-forest regions characterized by these processes. The development and deforestation in Morona Santiago resembles processes that have occurred in other rain-forest regions around the world. Development in Morona Santiago progressed through several recognizable stages like frontier development elsewhere (Bylund 1960; Hudson 1969; Brown, Sierra, and Southgate 1992:955). The successive periods of in-migration from outside the province and migration from west to east within the province correspond to the first two stages in models of frontier development. The farm abandonment on the fringes of the Upano—Palora region during the 1980s may mark the onset of consolidation, the third stage of frontier development.

Morona Santiago's deforestation, like its development, follows a general pattern observed elsewhere in the world. Growth coalitions appear to play an important role in processes of deforestation in other places with large forests, just as they do in Morona Santiago. Studies of deforestation in Southeast Asia, West Africa, and Latin America, describe the activities of growth coalitions, but the reports are more suggestive than conclusive because they do not focus on the coalitions' activities. The chief alternative form of land clearing in large forests involves lead institutions that open up regions for free riding peasants who clear land along newly constructed roads. Just how different are processes initiated by lead institutions and growth coalitions? While lead institutions and growth coalitions develop regions in quite different ways, regions opened up by lead institutions may, as they undergo settlement, give rise to growth coalitions that stimulate further deforestation.

If this historical sequence occurs frequently, then initial differences in land clearing between the two types of region will not persist, and the growth coalition model of deforestation will apply to a wider range of regions than our original discussion suggested. A brief case study of regional development initiated by a lead institution, in this instance an oil company in Ecuador's northern Oriente, addresses this question.

Missionaries established the first settlements in the northern Oriente much as they did in the southern Oriente. Carmelite Brothers in the Aguarico basin along the northern border with Colombia established missions that became the foci for settlement and land clearing by colonists from the Sierra. The priests did not have the economic resources necessary to build access roads into the missions, so few colonists followed the priests into the rain forests (Garcia 1985).

The pattern of regional development changed completely with the discovery of oil in the late 1960s. Between 1969 and 1972 a consortium of oil companies airlifted approximately 40,000 workers into the region to drill wells and build access roads; 90 percent of the workers entered the region with the intention of claiming a parcel of land along one of the many new roads.[1] They spent their days off inspecting unclaimed lands.

To file individual claims to land, workers had to join a state-mandated colonization cooperative. Composed of part-time colonists and absentee owners, most of the cooperatives proved to be "fantasmas"—only in exceptional cases did the members work together to clear land. One of the more honest part-time colonists admitted when he filed his claim that he did not have any tools to work the land![2] Most part-time colonists, like this person, sought land for speculative purposes—they had no interest in clearing it. In the first two years of settlement about one quarter of all of the claimed land in the region changed hands in speculative transactions. The second owners began to clear the land.[3]

A classic corridor-shaped pattern of deforestation emerged during the next ten years. Late-arriving colonists acquired land in the second, third, and fourth rows back from the roads. With long, narrow lots, 250 x 2000 meters, farmers in the last row had to walk at least six kilometers to reach the road. As in Morona Santiago the extent of cleared land declined sharply with distance from the road.[4] The colonists with claims to these interior lots faced the same set of disadvantages as the colonists with farms far from the road in Morona Santiago. Conflicts with colonists close to the road about rights of passage complicated the marketing of products from the interior farms, and governments refused to provide funds for the construction of feeder roads (Hiroaka and Yamamoto 1980:432; Gonzalez and Ortiz de Villalba 1976:21). Under these circumstances colonists cleared little land in the interior lots, and significant numbers of them began to abandon their farms during the mid-1970s (Barral, Oldeman, and Sourdet 1976:7).

These changes in the fortunes of settlers affected rates of deforestation. Rapid deforestation occurred between 1969 and 1973 when the oil companies constructed a network of roads, and colonists cleared large amounts of land along new roads. It slowed down after 1973 as arriving colonists claimed interior lots, which they developed slowly, if at all (Barral, Oldeman, and Sourdet 1976:7). The extent of cleared land continued to increase during the late 1970s because the oil companies extended their network of access roads into new areas, and colonists continued to settle along new roads. Several large enterprises—a cattle ranch, two oil palm plantations, and a logging company—also began to clear land during this period (Sandoval 1987:168–172).

For colonists with interior lots the efforts of local leaders offered some hope for the eventual construction of farm to market roads through their lands. Almost half of the colonists in the Lago Agrio region came from drought-afflicted regions in the highland province of Loja. A number of them formed groups and elected leaders prior to their arrival in the Oriente (Barral and Orrego 1978:6). When the leaders arrived in the Oriente, they acquired land, set themselves up as merchants, and took an intense interest in local politics.[5]

Without a municipal government or a representative in the provincial government, the leaders could not find politicians who would support their pleas for local initiatives. When the national government created new municipalities and eventually a new province in the late 1980s, it created political posts for colonist leaders, who could then provide funds for feeder roads in return for political support. By creating local governments, the central government created sources that coalitions could tap for funds. Modest increases in the construction of feeder roads and in rates of deforestation followed. The lead institution's initial efforts created conditions which, a decade later, gave rise to growth coalitions.

A similar sequence of events occurred among private sector producers along the Napo River. A lead institution, in this instance, an

African palm plantation, began to clear land along the Napo River in the mid-1970s. A few years later the company stopped clearing land and instituted a policy of indirect expansion. If small farmers with lands adjoining the plantations would promise to grow African palm and sell their harvests to the plantations, their owners would build feeder roads for the farmers and provide them with technical assistance in cultivating African palms (Guerrero 1987:246–249). Through this arrangement the small landowners got assistance in intensifying production and the plantation owners increased the size of their enterprises without becoming embroiled in land conflicts with neighboring peasants. The resulting expansion in African palm cultivation came at the expense of the surrounding rain forest. In this case, as in the oil fields, a lead institution initiated the land clearing, but growth coalitions made vital contributions to subsequent land clearing in the region.

Comparisons of deforestation in the northern and southern Oriente suggest that, while the processes differed dramatically at their inception, they converged over time. The speed with which the oil companies built roads in the northern Oriente compared with the slower moving efforts by the Salesians and CREA in Morona Santiago explains both the greater prevalence of land speculation and the faster pace of deforestation in the north.

These differences in the pattern of deforestation began to disappear after 1975 in large part because similar patterns of land distribution developed in the two regions. In both places smallholders acquired most of the land, and they sought alliances with powerful actors who could help them obtain improvements in infrastructure. This similarity suggests that one of the main differences in patterns of deforestation, between places opened up by lead institutions and places developed by growth coalitions, may be temporary. After an initial surge of deforestation started by either a lead institution or a growth coalition, further land clearing depends largely on the ability of local elites and smallholders to form new growth coalitions.

The degree of similarity between land use patterns in the northern and southern Oriente lends credence to the convergence thesis. Large landowners leave a higher proportion of their lands in forest in both regions (Ministerio de Agricultura/Pronareg 1980b:23–28).[6] The differences between colonists and Amerindian land use exist in both regions. Lowland Quichua in the north, like the Shuar in the south, clear a lower proportion of their lands and devote a higher percentage of their cleared land to the cultivation of crops than do the colonists. The colonists devote most of their cleared land (90%) to pasture for cattle (Ministerio de Agricultura/Pronareg 1980b:73; Barral and Orrego 1978:29). Like the Shuar, the Quichua cleared land in some locales at rapid rates during the early 1970s in an attempt to insure continued control over the land (MacDonald 1984). The more acculturated Quichua, like the acculturated Shuar, want to expand agricultural production for the market, and to that end they lobby for the construction of feeder roads into their villages.7 Like the Shuar, the lowland Quichua have in recent years migrated in large numbers to areas being opened up for settlement and deforestation. Large numbers of lowland Quichua have moved from the Tena–Puyo region to the oil rich Lago Agrio region in order to work for the oil companies and acquire land (Barral and Orrego 1978:27). In sum, the ethnic differentials in rates of deforestation in the southern Oriente also characterize the northern Oriente.

The comparison of places opened up by lead institutions and growth coalitions suggests divergence in land clearing patterns during the initial period of development and convergence in subsequent periods. Although the land clearing patterns may converge over time, the initial differences in the organization of regional development may continue to shape land clearing because lead institutions and growth coalitions respond so differently to political pressures to preserve rain forests. The success of policies to prevent further deforestation will vary from region to region, and it depends in part on the type of organization that drives deforestation in a place.

POLICY IMPLICATIONS

To construct a forward-looking argument about policies from a backward-looking empirical analysis requires an appreciation for recent changes in the context in which policymakers work. In the case of tropical deforestation the context has changed dramatically. Tropical deforestation has evolved from a relatively obscure land use issue into a cause celebre in the past fifteen years. The growing salience of the

Table 8.2. Policies to Reduce Tropical Deforestation

Policy	Extent of Protected Area	Incidence of Political Conflict	Policy Approach	Policy Instrument
Company enclosure	Small	High	Direct	Law[a]
Park creation	Small	High	Direct	Law[a]
Indigenous reserves	Small	High	Direct	Law[a]
Extractive reserves	Small	High	Direct	Law[a]
Agroforestry	Small	Moderate	Direct and indirect	Law and Economics[a]
Social forestry	Small	Moderate	Direct and indirect	Law and Economics[a]
Agricultural intensif.	Large	Low	Indirect	Economics
Integrated rural devel. /land reform	Large	Low[b]	Indirect	Economics
Urbanization	Large	Low	Indirect	Economics
Population control	Large	Low	Indirect	Economics

[a] These policies would be components in systems of agroecological zoning established by central governments.

[b] Integrated rural development programs would generate intense political conflict if they include a land reform.

issue has inspired numerous proposals for policies to slow down or stop deforestation in particular locales (Anderson 1990). The following description of these policies includes an attempt to gauge their effectiveness in slowing deforestation. Table 8.2 outlines the policies, their attributes, and some of their effects.

Recent analyses of development in the Amazon basin have emphasized how the groups interested in rain forest destruction and preservation vary from region to region. These diverse political ecologies imply that no single policy will slow or reverse rates of deforestation. Only a set of policies, each with its own variable effects across locales, can produce dramatic declines in rates of deforestation. The policies listed in table 8.2 have either direct or indirect effects on deforestation. Policies with direct effects create groups that defend the forests against destruction by other groups. The defenders come in a wide variety of guises: park rangers, company guards, indigenous peoples, and small farmers practicing agroforestry. These policies reduce the supply of unprotected forests. Policies with indirect effects discourage rain forest destruction by directing the entrepreneurial energies of forest destroyers elsewhere. Smallholders may decide to invest their savings in a truck to transport crops rather than sink the money into a scheme to acquire additional land farther into the forest. These policies reduce the demand for deforestation. In the following pages we describe the two sets of policies. We then discuss their relative effectiveness in preserving large blocks and small islands of forest.

Increasing the Number of Protected Forests

Until the mid-1980s oil and mineral companies built penetration roads into rain forest regions without any concern for incidental deforestation along the new roads. Under political pressure from nongovernmental organizations (NGOs) concerned with the environment and indigenous rights, companies have recently begun to enclose the areas in which they work in order to prevent spontaneous colonists from settling along company roads.

For example, CONOCO, an American oil company, recently received permission to open up a new oil field in a national park in eastern Ecuador. To service the wells, it will build a penetration road into the park and an Amerindian reserve. To prevent deforestation along the road, the company will build a fence around the area, restrict entry by road, and use satellite images to monitor land use within the concession.[8] In effect CONOCO has responded to pressure from NGO's by enclosing the region within which it works. Through enclosure it hopes to minimize deforestation and damage to the local Amerindian group, thereby eliminating two sources of political controversy about its work. In another example the Companhia Vale do Rio Doce (CVRD) has claimed a large area around its Carajas mine and established a nature preserve in this area (Hall 1989:166).

Like large corporate enterprises in developed countries, CVRD and CONOCO have narrow interests in a particular economic activity, and they do not oppose land use restrictions as long as they permit the activity, like oil extraction, in which they engage (Walker and Heiman 1981). Where small capitalists in coalition with lead institutions clear most of the land, environmental agencies and NGOs encounter more opposition. For example, near Brazil's Carajas mine small businessmen fire their smelters with charcoal made from the surrounding forests.[9] The associated charcoal industry clears thousands of hectares per year, and it is unlikely that environmental groups can do more than establish some outer limits on the size of the deforested areas. In sum they can control the extent of the deforestation, but they cannot stop it.

Other policies also work through the force of law. One set of policies, promoted by conservation organizations like the World Wildlife Fund, puts an emphasis on preservation; it stresses the importance of creating parks and indigenous preserves to protect the forests and their peoples. A related set of policies focuses on strengthening political and economic pressure groups that will defend the forests. Through the creation of extractive reserves for forest product collectors, these analysts hope to strengthen groups with vested interests in the preservation of the forests who will oppose land clearing schemes by companies or colonists (Schwartzman 1989; Hecht 1989:233). Local residents extract a wide variety of forest products, including fruits, rattan, timber, and medicinal plants from the reserves. The extractive reserve scheme with the widest application involves the sustained yield management of tropical forests through the logging of narrow strips of forest (Hartshorn 1989).

None of these policies promise to change the economic rewards of starting a farm in the forest. Advocates of parks, extractive reserves, and Amerindian reservations argue that peasants and ranchers can be stopped by drawing a boundary line in the forest. In practice countries find it difficult to maintain protected areas. Park creation requires, to be effective, widespread public support for the creation of parks and a well funded park service to police the parks (Schwartzman 1989; Ramos 1988:430–431). Invasions occur because the people who live around the parks often regard the untapped resources within park boundaries as opportunities to be exploited, and park services do not receive enough funds to police their boundaries (Foresta 1991; Gildesgame 1989).

Not surprisingly, the proponents of parks acknowledge the importance of political conflict in the preservation of rain forests. Policies emphasizing reserves create controversies because they allocate large areas of tropical forest to small numbers of people. Rubber tappers in Brazil, indigenous peoples in wildlife production preserves in Ecuador, and logging companies in Ecuador and Indonesia have all faced incursions and public protests by poor outsiders who want to make more intensive use of the resources, in most cases through agriculture (Schwartzman 1989; Nations and Hinojosa 1989; Barral 1979; Peluso 1990).[10] By strengthening pro-conservationist forces who struggle against pro-development groups, these policies conserve rain forests.

Agroforestry represents a somewhat different approach to the problem because conceivably it could alter the economic incentives facing the large populations who convert so many forests into fields. The most common type of agroforestry involves peasants who cultivate tree crops in association with row crops (Nair 1990). The most environmentally benign agroforestry regimes only cultivate tree crops, but few farmers practice this type of agroforestry. Those who do face severe constraints in marketing their products. They produce small quantities of a large array of products and find it difficult to market these goods (Padoch and de Jong 1989:108–110). Markets for all of these products only exist in close proximity to large numbers of urban consumers, and even then the small volume of each item makes them expensive to collect and ship to market (von Thunen 1851).[11] Farmers who practice agroforestry in Peru have reduced their marketing problems by specializing in the production of a single crop like cacao. Other farmers avoid this problem by treating agroforestry as a supplementary source of income and devoting a large portion of their land to pasture or row crops (Nair 1990). While even limited forms of agroforestry provide environmental benefits, only the pure forms serve as substitutes for the rain forest, and the constraints noted above make it unlikely that these systems will cover large areas.

Like agroforestry programs, social forestry policies only provide protection for relatively small areas of forest. By assigning rights of use and responsibilities for protecting forests to nearby villagers, social forestry programs promise to protect islands of rain forest in densely settled areas. The salience of these schemes in discussions of forest preservation in South and Southeast Asia reflects the prevalence of forest remnants in this region.[12] To succeed, social forestry requires fairly high population densities relative to the size of the conserved areas. Enforcement problems make social forestry programs unworkable in places with small populations and large blocks of forest like the Amazon basin.

Taken together, the parks, reserves, company enclosures, and agroforestry projects fit into schemes of agroecological zoning that the central governments in developing countries have elaborated at various times during the past ten years. These plans delimit the protected areas and channel future development initiatives into other areas. A lead institution intent on developing a protected area would have to negotiate a series of safeguards which would limit the damage done to the surrounding rain forest. International environmental groups have recently tried to augment the emerging patchwork quilt of protected areas with proposals for more centralized control of the Tropical Forest Action Plan (TFAP) and an international convention that would require countries to preserve a certain portion of their forests.[13] The groups' advocacy of "top down" regulation, coupled with their support of grassroots environmental organizations, suggests that the groups' primary commitment is to forest conservation through political control, either from the grassroots through local organizations or from the top through government regulators.

The limitations of the direct approach to rain-forest conservation become apparent when the size of the protected areas is compared with the size of the forests at risk. Although developing countries created many parks during the 1980s and conservationists invented promising new means for creating parks through debt for nature swaps, parks currently protect less than 5 percent of all tropical forests.[14] States in poor societies will not give up their rights to exploit vast unprotected areas, so it is unrealistic to expect that the bulk of the world's tropical forests will end up inside preserves. Even if countries could establish preserves, they can not afford to staff the park services needed to protect the preserves (Protang and Thomas 1990).

While most lead institutions acknowledge and to some degree respect agroecological zoning laws, local growth coalitions oppose the laws. The different responses to the law reflect differences in their interests in development. While lead institutions have a strong interest in a particular productive process, which may or may not be reconcilable with extensive forest cover, the members of growth coalitions have more general interests in development. They want to intensify land use in whatever way a developer sees fit and reap profits from the sale of land. By restricting land uses, zoning schemes limit the available options for intensification and work against the financial interests of coalition members. Their opposition to agroecological zoning underscores that these policies rely on coercion to accomplish their ends. The policies do nothing to diminish the economic incentives to form growth coalitions. Only in places where indigenous groups or others stand ready to defend a preserve do these plans discourage growth coalitions, and then the chief effects of the restrictions may be to channel the energies of the coalitions toward other, uncontested tracts of rain forest. Clearly these policies will not stop tropical deforestation by themselves.

Reducing the Demand for Deforestation

Recent discussions of policies to reduce rates of tropical deforestation have generally ignored indirect policy approaches. These policies work through economic incentives, either making it less rewarding to clear forests or more rewarding to invest in already deforested areas. In the best-known application of this approach the World Bank cut off funding for penetration roads into rain-forest regions after environmentalists established that bank financed roads encouraged massive deforestation in Brazil during the 1970s and 1980s. The much-maligned Tropical Forest Action Plan argues in a limited way for this approach. The plan acknowledges the importance of agroforestry initiatives, but above all it stresses the effectiveness of agricultural intensification and expanded production from tree plantations in reducing rates of deforestation (World Resources Institute 1985; Repetto and Gillis 1988).

The authors of TFAP recognize that smallholders and entrepreneurs will continue to exploit the forests and that political conflict with indigenous peoples and forest product collectors will save only a portion of the forests, so they argue for an additional set of policies. They advocate lowering the price of tropical commodities through the intensified production of wood and agricultural commodities. Eventually prices will decline to the point where logging primary forests or converting them into fields would not produce profits for colonists, loggers, or their financial partners (Spears 1986:394; World Resources Institute 1985 1:30–33). The problems with TFAP in its first five years have more to do with the way foresters have implemented the plan than with its conceptual underpinnings. The foresters have placed an undue emphasis on the expansion of forest plantations, and they have neglected other, important components of the plan, like agricultural intensification (Winterbottom 1990:5).

Prescriptions for agricultural intensification in the absence of other policies, usually benefit capital intensive producers and drive marginal producers out of the market as the better-financed producers adopt innovations that increase the size of their harvests. With prices for agricultural commodities falling, small farmers on the margins of the forest would scrap plans for the conversion of more forest into fields, and rates of deforestation would probably decline.

Would the small farmers actually leave the rain forest? In the absence of other economic opportunities, the small farmers' search for a livelihood might approach the pattern observed among the second generation of settlers in Sinai. Overproduction of narangilla brought a steep decline in the price of the fruit beginning in 1986. Peasants responded by reducing the rate of forest clearing and initiating other plans for earning a livelihood. Many of these plans involved migration to urban areas. Virtually everyone in Sinai counts a rural to urban migrant among their close friends, so chain migration to an urban place is a realistic option for them.

Other plans involved the creation of growth coalitions to claim and clear forested land, but only a small proportion of these coalitions actually succeeded in acquiring land and developing it. The decline in the price of narangilla reduced the rate of deforestation around Sinai, but it did not prevent the formation of growth coalitions. Rather it accelerated the peasants' search for all types of economic opportunities. Families hatched plans to make money at a greater rate than they had five years earlier, and some of these plans involved the acquisition and clearing of additional rain forest. In sum agricultural intensification with attendant price declines would reduce rates of deforestation on existing farms, but otherwise it has diffuse short-term effects. Declines in the prices of agricultural commodities, coupled with increased economic opportunities in urban places and long-settled rural areas, would produce long-term declines in deforestation.

Integrated rural development (IRD) programs would complement the agricultural intensification policies (Sawyer 1990). IRD programs would target the sending regions for colonists and attempt to strengthen village economies in ways that would enable them to retain their populations rather than sending them to new zones of settlement. Some IRD programs would include a land reform.

The impact of a land reform on tropical deforestation depends on regional patterns of rural–rural migration and regional variations in the application of the law. If the regions sending migrants to the frontier have little forest and a skewed distribution of land, a reform might help these places retain larger proportions of their populations, thereby reducing the stream of forest destroying migrants to the frontier. Otherwise, a land reform would have little effect. For example, Ecuador has a highly skewed distribution of land, but almost all of the migrants to the rain-forest regions have come from places with relatively equal distributions of land. In the 1950s and 1960s the migrants came from communities of smallholders in the southern highlands; in the 1980s they came from recently settled colonization zones where middle-sized properties, 15 to 50 hectares, predominate. As noted earlier, most of these spontaneous colonists have some financial backing from their families. A land reform would have no impact on migratory streams originating in these places because the law would not redistribute land in these communities.

The effects of a reform on rates of deforestation often depend on the applicability of the provisions to tropical zones. Many reforms in Latin America single out underutilized *latifundia* for expropriation by requiring that landowners work a certain proportion of their land or face expropriation. Applied in a tropical zone, this type of law increases rates of deforestation by forcing landowners with a large proportion of their land in forest to clear extensive proportions of it in order to avoid expropriation. The passage of the 1973 agrarian reform law in Ecuador provoked this type of reaction in the Oriente. Afraid of expropriation, Shuar and lowland Quichua landowners cleared extensive amounts of land during the mid-1970s (MacDonald 1984). In sum, under a set of highly specific conditions a land reform can slow down deforestation; in other circumstances it has the reverse effect or little effect at all.

Because the sending regions are often areas of relatively recent settlement, some of these IRD programs would target settled places near large blocks of rain forest. The programs would be selective in their programmatic foci, emphasizing the generation of employment opportunities outside of agriculture and improvement in extension services concerned with agricultural intensification. The importance of adequate extension services in facilitating agricultural intensification cannot be overestimated. With encouragement from extension agents the Shuars' move into cash cropping could result in the widespread adoption of agroforestry practices. Other acculturated indigenous groups in Asia, Africa, and Latin America have the same potential.

Because rapidly growing populations create pressures to clear islands of forest in settled areas and provide recruits for the coalitions that invade forested areas, declines in fertility should over time contribute to lower rates of deforestation (Rudel 1989a:331). To induce declines in fertility, IRD programs should focus on education and employment for women as well as men. When IRD planners decide to build roads, they should locate them in already settled areas far from major blocks of rain forest.

The aim of the IRD programs is not to prevent the formation of growth coalitions but to divert their energies toward economic development projects that do not involve the clearing of additional tracts of rain forest. A similar strategy of directing development initiatives at urban places would reinforce the effects of the IRD programs. By employing young, landless peasants, labor-intensive economic projects in urban places would siphon off the poorer laborers whom the wealthier coalition members recruit to do the hard physical labor of opening up the forests for settlement and deforestation. More generally, these indirect approaches to forest conservation recognize that the problems of tropical deforestation are inextricably linked to processes of development. To reduce deforestation, policies must redirect development toward more environmentally benign activities.

This emphasis on changing the structure of opportunities follows from the theory proposed here. Like the proletarianization thesis, the lead institution/growth coalition theory emphasizes political economic structures, but it focuses on a structure of opportunities rather than a structure of class domination. It disputes the empirical assertion by proletarianization theorists that an impoverished and oppressed peasantry represents an important source of tropical deforestation in places with large blocks of rain forest. Ventures by lead institutions and growth coalitions cause most deforestation in these places, and peasants sometimes play important roles in the land clearing. Under these circumstances policies to reduce deforestation must alter the array of opportunities available to wealthy investors as well as peasants. Urban development and geographically focused rural development promise this type of change in the structure of economic opportunities.

Policies for Large and Small Forests

Only packages of policies that increase protected areas and reduce the attractiveness of clearing land can produce significant declines in tropical deforestation. The effectiveness of a policy package will probably vary from places with large blocks of forest to places with small islands of forest. In places like the Amazon basin the direct approach cannot secure more than 15 to 20 percent of the forest. In this setting indirect approaches that redirect entrepreneurial energies elsewhere offer the best hope for preserving the bulk of the forest. By reducing the returns to pioneering and increasing the returns to economic development in urban and long-settled rural areas, these policies preserve the forests by steering entrepreneurially minded individuals toward settled rather than forested areas. A package of these policies would have an impact on all of the forests in the Amazon basin, so they hold the most promise for slowing deforestation over a large area. These indirect policies should form the cornerstone of plans to conserve large blocks of forest in Africa, Asia, and Latin America.

In places like Costa Rica that have small forests, encroachment by an expanding smallholder population could eliminate the rain forest. In this setting a combination of direct approaches in the form of expanded preserves, social forestry, and agroforestry, might protect most of the country's remaining rain forests. Indirect policies like agricultural intensification might reduce the human pressure on the few unprotected forests, but their overall impact on forest cover would be slight because so few forests would remain unprotected. In places with small forests, direct approaches with their emphasis on protection deserve the highest priority.

The policies outlined here, like most prescriptions for change, may fall on deaf ears. In this instance the suggested policies may be more difficult to ignore because they build on shifts in the spatial distribution of populations already under way. Despite the recent expansion of agriculture into tropical forests throughout the third world, much larger numbers of peasants have moved to cities than to agricultural frontiers since 1970 (Vining 1986). Urban development projects would provide jobs for these workers and induce further rural to urban migration from the colonists' places of origin. In rural places well served by roads, nonagricultural employment has grown rapidly in recent years, allowing people to "leave the land but not the village" (Veeck and Pannell 1989; Rudel and Richards 1990). IRD programs would expand the number of rural places able to retain their populations. In this sense the mix of policies outlined here reinforces already existing trends in tropical regions.

This set of policies should slow but not stop tropical deforestation. Local governments will undoubtedly use some IRD funds to free up money for use in ventures to open up additional tracts of rain forest. Prosperous villagers, poor villagers, and local elites will continue to form growth coalitions to exploit economic opportunities, but hopefully a smaller number of these opportunities will involve the destruction of primary rain forests. In addition, these policies have a social cost. Falling commodity prices bring economic ruin to small farmers working poor tropical soils in marginal places, so they must find a new livelihood. Urban economic growth holds out the promise of more environmentally benign employment for these people, but the adjustment to new work in new places occurs slowly and painfully, if at all.

Agricultural intensification and urban growth will not reduce rates of tropical deforestation overnight. The commitment to agricultural research and extension inherent in programs of intensification takes years to implement, and variable amounts of time must elapse before the increases in land productivity begin to reduce the demand for additional agricultural land. In contrast, the direct approach to forest conservation, by walling off areas, could produce short-run declines in tropical deforestation. In this respect as well as in others, the indirect approach to forest conservation complements the direct approach. The optimal combination of approaches should vary, along the lines outlined above, with the size of the forests in a region. This untidy mix of policies has the best chance of preventing the destruction of the world's remaining tropical rain forests.

APPENDIX A
Research Methods

■■
■■

Thinking about tropical deforestation leads quickly to the conclusion that some of its causes originate in rural communities and others stem from policies and decisions made in urban centers of power and commerce. To incorporate influences from the local community and distant political economic centers into a single explanation, we needed ethnographic data on peasants in rural communities and historical data on elites in regional and national centers of power. Accordingly, our research combined a study of smallholders' decisions to clear land with an investigation of the relevant aspects of the larger political economic order (Vayda 1983). To ground our analysis in the thinking of people who clear land, we carried out a comparative study of small landowners in two recently deforested places. To understand the larger political economic forces which affected rates of deforestation, we searched through institutional archives and interviewed key informants in agencies. Both the interviews and the archival research provided information which enriched our interpretation of patterns discovered through a third type of research, a remote sensing analysis of forest cover in Morona Santiago. The details of this three pronged research strategy are outlined below.

COLLECTING DATA IN THE FIELD: SINAI AND UUNT CHIWIAS

Research Strategy

The Upano—Palora region of Morona Santiago had two distinct advantages over other areas as a site for field research on tropical deforestation. First, tremendous amounts of land clearing occurred in this region during the period immediately preceding this study. Other areas like the Upano valley had undergone tropical deforestation more then thirty years earlier, so participants would have been hard to find, and satellite imagery would not have been available for the period during which most of the deforestation occurred. In other places, such as the Morona–Huasaga watershed, extensive land clearing had just begun, so the full outlines of the process were not yet apparent. Given the magnitude and the timing of deforestation in the Upano—Palora region, a field study in this area promised to yield a rich lode of information about the dynamics of the process. Because Tom Rudel had worked in the Upano—Palora region as a Peace Corps volunteer and knew many of the colonists, gaining access to the colonists in this zone did not present a problem.

Access to the Shuar landholders did present a problem. Because both mestizo colonists and the Shuar own substantial amounts of arable land in western Morona Santiago, a comprehensive study of deforestation in the region required fieldwork among both the colonists and the Shuar. A language barrier and long-established hostility between the two groups complicated the task. Having worked with CREA's colonists, Rudel would have had a difficult time getting permission to conduct research among the Shuar. This problem was resolved by bringing Bruce Horowitz into the project. Horowitz had worked with the Shuar as a Peace Corps volunteer, and his participation made it possible to divide up the field work by the ethnicity of the group under study. Horowitz did the field research in the Shuar *centro* of Uunt Chiwias, and Rudel did the field research in the colonist community of Sinai.

To carry out the field research, we had to secure the permission of a number of authorities. In addition to securing permission for the study from the federal government's Instituto de Patrimonial Cultural, we asked a number of Shuar authorities for permission to carry out the study. At different times the President of the Federacion de Centros Shuar, the President of the Chiguaza Association of Centros Shuar, and the *sindico* (mayor) in the Shuar *centro* of Uunt Chiwais approved the study after we explained its aims and methods to them. Undoubtedly, Mr. Horowitz's extensive experience with the Shuar during the 1970s and early 1980s proved important in securing permission to carry out the study.

Our firsthand experience with the people and the subject of investigation is unusual for a social scientific study and requires some comment. Problems of adopting the respondent's viewpoint and "going native" usually increase among researchers who know their respondents well. These problems seemed manageable to us in part because our respondents did not have well-developed opinions about tropical deforestation. They all participated in the process and regarded it with some ambivalence. We described the focus of the investigation to all of our respondents and never encountered any hostility to the aims of the research.

Our prior experience in the region did have two distinct advantages. First, many students of tropical deforestation spend little time in the jungle clearing land and much time in new frontier towns interviewing people about their experiences in the jungle. In other words these studies rely very little on participant observation and very much on retrospective accounts of events. Unlike most researchers we were participants in some of the events that we describe, but we did not keep detailed notes of events as they occurred, so our experience cannot be construed as participant observation. The material from these experiences can best be described as memoirs. Both of us worked in the rain forest, helping people carve up and clear land. In 1969 and 1970 Rudel explored the area around Sinai with the colonists, surveyed land, and helped the colonists plan their village. His experiences form the basis for much of the analysis in chapter 5. Horowitz surveyed land from 1970 to 1973 and again in 1982 in a large number of Shuar *centros*, including Uunt Chiwias.

To check our own recollections of events, we asked our respondents for their accounts of the same events. In the same conversations our respondents would often provide valuable updates on stories that began with the first settlement and continue to the present day. Second, because we both played prominent but disinterested roles in the early histories of the two communities, we had little trouble gaining the confidence of the people we studied. In many instances we interviewed people with whom we had worked fifteen years earlier, surveying land or organizing a community. This fund of common experience between researcher and respondent made it easier to engage in frank discussions about the successes and failures of the ensuing fifteen years.

Our research strategy has been iterative rather than linear in its execution. We have opted for multiple visits to our research sites over a long period of time rather than one long stay. Rudel collected data on trips to the Oriente in 1971, 1980, 1985, 1986, and 1989, spending a total of ten months in the field. The first visit, in 1971, included a three-week stay in the oil field regions of the northern Oriente. The transcripts from forty interviews during this period with colonists, government officials, and oil company employees form the basis for the comparative case study presented in chapter 8. Horowitz, a resident of Ecuador, did the fieldwork in Uunt Chiwias during the summer of 1986; since then he has talked with key Shuar and Salesian informants whenever we needed the answer to a particular question.

This research schedule has made it easy to incorporate a longitudinal dimension into our research. On each trip we would get updates on the efforts to open up particular areas for development and deforestation. Our analysis of the remote sensing data also benefited from this back-and-forth schedule. Following an extended period of field research (five months) in 1986 when we conducted the survey in Sinai and Uunt Chiwias, we began the remote sensing analysis in the United States. In 1987 Horowitz came to the United States to participate in the remote sensing analysis. In 1989 armed with questions about patterns of deforestation evident in the satellite images, but inexplicable in terms of our field data, we returned to the field and asked additional questions about patterns of forest destruction in particular locales that we had overlooked during our earlier field research.

The Conduct of the Survey

Cultural differences between the colonists and the Shuar led to some differences in the conduct of interviews. The interviews with the Shuar were done in the presence of the village's *sindico*, while the interviews with the colonists were not done with anyone other than family members present. Two contextual factors account for the *sindico*'s presence. First, because the Shuar in Uunt Chiwias live in houses dispersed throughout the *centro*, Horowitz could have easily lost his way going from house to house, so the *sindico* served as a guide. Second, the interviews were conducted in a mixture of Shuar and Spanish. When Horowitz's Shuar or the respondent's Spanish failed, the *sindico* assisted with a translation. Horowitz's understanding of Shuar was sufficiently good so that he is sure that the *sindico* always communicated the proper meaning of a question in his translation into Shuar.

The interviews with the Shuar followed a pattern. Horowitz and the *sindico* would approach the house, announce that they were coming, and be invited into the house. The respondent would sit at one end of the house, and Horowitz would sit at the other end or along a side wall. As is customary, Horowitz would wait in silence for some period of time until the head of the household (the respondent) decided to speak. Then the household head would converse for a period of time, between ten and forty-five minutes, with the *sindico* about the news of the *centro* and other items of interest. Next, he would ask the interviewer about the project. In about half of the houses Horowitz was asked to sing one of the Shuar songs that he and the villagers of Uunt Chiwias had sung when they worked together. During this time they were served *chicha* (manioc beer) and usually a meal. After the meal Horowitz would begin the interview. The actual interviews took relatively little time, less than one hour in all cases. In total Horowitz interviewed thirty-six out of the forty-five landowners in the *centro*. The persons he failed to interview were not at home during the period in which he did the field research.

The obvious question raised by these procedures concerns the possible biasing effects that the presence of the *sindico* at the interviews may have had on the respondents' answers. The traditional independence of the Shuar as individuals and the conduct of the interviews in the respondent's home, where in Shuar culture a person tends to be most expansive, argues against any pronounced biasing effect. If some bias were to occur, it would most likely show up in responses to questions asking about opinions or plans for the future. Because most of the analysis focuses on matters of fact, the amount of forest cleared, for example, it should be relatively free of biases introduced by our interviewing procedures among the Shuar.

For our interviews with mestizo colonists in Sinai, we attempted to reach the owners of tracts of land that had been settled in the late 1960s and early 1970s. In practice this meant interviewing all of the landowners in the area surrounding the village. Some residents in Sinai arrived in the mid-1970s and received 30-hectare lots across the Ambusha, about a three-hour's walk from the village center. These late arrivals received smaller lots at later dates in more isolated settings than the Shuar in Uunt Chiwias. For these reasons the late arrivals would not have been an appropriate comparison group, so we did not interview them.

The colonists in the immediate vicinity of the village center settled in the area about the same time as the Shuar organized their community in Uunt Chiwias. These colonists received similar-sized tracts of land as the Shuar. The government built roads through both communities during the 1970s, so access to markets does not vary much between the two places. In effect, by choosing the Shuar in Uunt Chiwais and the colonists close to the village center in Sinai as comparison groups, we matched the two subsamples on all but one important attribute — ethnicity. In the lexicon of quasi-experimental social science this is a post-test only comparison group design (Campbell 1988:219).

We interviewed thirty out of the thirty-three landowners close to the village of Sinai. On repeated occasions we tried without success to reach the remaining three landowners.[1] Because most of the colonists spend their days working on their farms, we conducted almost all of the interviews during the evenings or on weekends when the colonists could be found in their homes. On many occasions the discussions began with a recounting of the major events in our personal lives since our earlier work together. After explaining the purpose of the study, the interviews began; they ranged in length from forty-five minutes to two hours in length. Some respondents had participated in several attempts to acquire new lands, and they gave sometimes lengthy accounts of the events surrounding these episodes. These stories and questions about them made some of the interviews much longer than others. In some cases where ambiguities remained after the interview I would return to the respondent's house several days later to ask a clarifying question.

In addition to the formal interviews we relied on key informants in the village for all types of information. One informant became a research assistant who helped us conduct a census of the households in Sinai. A number of people spent innumerable evenings on the veranda after dinner talking about the history of the community and its future. They also provided useful checks on the accuracy of stories about land acquisitions that would come up during interviews. In many instances one informant would tell us about a failed expedition to acquire lands and, by raising the subject with other informants who participated in this venture, we would get multiple stories which we could piece together to get a complete account of the expedition.

The Survey Instrument

The questions in the survey focused on three aspects of the respondent's life: (1) the development and deforestation of his land; (2) the acquisition of additional land; and (3) the economic and demographic characteristics of his household. We developed first drafts of the interview schedule after a visit to Ecuador in 1985 and modified it after some pretesting in 1986. Brief descriptions of the questions are presented below in the order in which they were asked.

The History of Land Use. The interview began with a general question about the major problem which the family faced at the time of the interview. The following set of questions focused on the history of land use on the parcels of land which the respondent owned.[2] First, we asked how much land he or she owned.[3] Then we asked him to describe the location of the parcels and to indicate the quantity of land in each parcel used for pasture, narangilla, and other crops. From these figures we could calculate the proportion of the respondent's land that remained in forest. We also asked for a count of the livestock owned by the respondent. Among the Shuar we asked for an estimate of the number of cattle they had sold in the previous year. We then asked a series of questions about the history of land use on their properties. Did they begin clearing forest for pasture as soon as they settled on the land? What role has the cultivation of narangilla and other cash crops played in decisions to clear land? We also asked about the location of the respondent's land and how he had acquired it. If there were several parcels, we would ask him to tell us how he acquired each parcel.

We then asked a series of questions about the land clearing, beginning with a question about the lands still in forest. Why had they been left in forest? Were the lands left in forest extraordinary in some topographical or geographical sense? Who did most of the work in clearing the land? We then listed a number of factors and asked the respondent to indicate if any of these factors had influenced his decisions to clear land. The list included the following factors: the acquisition of title to the land, loans from either banks or individuals, changes in the prices of products, changes in the health of family members, and a residual category for all other factors. When a respondent mentioned that a particular factor had proved influential, we followed up with questions about the ways in which this factor had influenced his decisions to clear land. To pick up any plans for the future that might include agricultural expansion, we asked the respondents how they planned to provide for their children.

The Acquisition of Additional Lands. This section began with a question asking the respondent to describe his various occupations in order of importance. We then asked the respondent if he wanted to acquire additional lands for cultivation. We asked if he had plans to acquire additional lands and, if he did, what those plans were. We went on to ask about past attempts, successful and unsuccessful, to acquire additional lands for farming. We asked these questions in an open-ended way in an attempt to elicit accounts of land acquisition schemes. Once the respondent began to recount the events surrounding a particular attempt to acquire land, we would ask a series of specific questions in order to get the respondent to expand on his account of the events. If a person had not attempted to acquire additional lands, we asked him why he had chosen not to acquire new lands.

Personal and Household Characteristics. We then asked the respondent a series of questions about himself and his family, beginning with a question about the family's length of residence in the community. Several other questions about residence in other places followed. Through these questions we hoped to pick up any temporary labor migration or prolonged residence in a community of origin in the highlands that might have affected land clearing patterns in the rain forest. We then asked about the respondent's age and the number of years he attended school. The next question asked about the composition of the respondent's household, the number of persons who lived in the household, their age, sex, and relation to the household head.

The final questions focused on household income. We asked the respondent to list the household's sources of income, and the percentage the household's income that came from the largest single source of income. Using procedures described in chapter 7, we estimated the household's annual income from these data. The interview concluded with a general question about the future. We asked the respondent what worried him most about his family in the future, and how he planned to deal with this problem.

ANALYZING SATELLITE IMAGES

To measure variations in the rate and pattern of deforestation over an extended area of the Ecuadorian Amazon, we carried out a series of remote sensing analyses on a LANDSAT multi-spectral scanner (MSS) image of southeastern Ecuador taken in July 1983. Normally, one would choose to use thematic mapper (TM) images rather than MSS images because the TM images have higher resolution. The minimum unit area in a TM image is 30 meters square as opposed to 80 meters square in a MSS image.[4] We purchased both MSS and TM images for the mid-1980s. The superior quality of the 1983 MSS image more than offset any advantages in resolution that we might have gotten from using the smaller unit area in a 1984 TM image, so we relied on the MSS image for most of the remote sensing analyses reported here. We could not get any LANDSAT images for a later date because they all contained extensive cloud cover, making them useless for a study of deforestation.

Reliance on the 80-meter minimum resolution almost inevitably means that our remote sensing analyses have misclassified some small cleared areas as forest. Close examinations of areas in the image that we know well from extensive on-ground experience indicate that all pastures and larger gardens around Shuar homes show up as cleared land in the image. Because the remote sensing analyses probably misclassify some of the smaller gardens as forest, our remote sensing estimates of deforestation in Shuar communities may be slightly lower than estimates derived from ground surveys.

We purchased MSS images from 1973 as well as 1983 and considered doing a longitudinal analysis of deforestation for the 1973–1983 period in Morona Santiago. Eventually we discarded this plan because the considerable technical difficulties of carrying out the analysis appeared to outweigh the increased understanding that might result from the analysis.[5] The decision not to carry out longitudinal analysis rested on the assumption that in our primary zone of interest, the Upano—Palora region, there was virtually no cleared land before 1960. The cross-sectional differences between colonists and Shuar in the extent of cleared land in 1983 represent the cumulative amount of forest clearing carried out by colonists and the Shuar between 1960 and 1983 because very few landowners allowed land to revert to forest during this period. Under these circumstances the gross differences between places in the amount of deforestation represent differences in rates of deforestation. Using the 1983 image, we carried out a fine-grained, cross-sectional analysis of differences between communities in the extent of deforestation. As explained in the endnotes that support the text, these analyses could not have been done without two additional sources of information. Maps from the Federacion de Centros Shuar made it possible to establish the boundaries of each Shuar *centro* on the satellite image. Personal knowledge of the boundaries of the colonist settlements in the Upano—Palora region made it possible to define these boundaries with reasonable accuracy.

RESEARCHING THE HISTORY OF REGIONAL DEVELOPMENT

In additional to published historical studies, we relied on three archives for information about the ways in which regional and national institutions influenced the pace of rain forest destruction in the Ecuadorian Amazon. The Salesian Mission in Ecuador maintains an archive of documents, studies, and reports on the Shuar and Morona Santiago, dating from the 1920s, which proved useful in reconstructing the early history of the region. The archives of CREA's colonization program provided valuable information about the history of the settlements it sponsored and the policies it pursued. Finally, for the past nine years the Salesians have published, semiannually, a compilation of press reports about indigenous peoples and the Amazon region. Each volume includes all of the articles concerning indigenous peoples and the Amazon region that appeared in Sierra newspapers during the previous six months. An unusual journalistic practice enhanced the value of the Salesians' press clipping service. Quito's oldest newspaper, *El Comercio*, has a practice of reprinting in reduced form their lead stories from their editions of twenty-five, fifty, and seventy-five years ago; those stories concerning the Amazon basin found their way into the Salesians' volumes. The occasional story concerning the Amazon region from fifty or seventy-five years ago made it much easier to follow the shifting course of the country's policies toward the region.

To supplement our archival research, we conducted thirty interviews with functionaries who had detailed knowledge of important events in the history of Morona Santiago's development. Most of these people were either Salesian priests, officials at the Federacion de Centros Shuar, or employees of CREA. Some of these interviews lasted only thirty minutes. The longest interviews, with the topographers at the Federacion de Centros Shuar, exceeded three hours in length. Invariably, we went to these people with specific questions about the development of the zone, about, for example, CREA's road building policies or the role of the Salesians in the creation of Shuar *centros*. By focusing on specific examples of road building or land acquisition, these informants frequently bridged the gap between general accounts of policies presented in the press and peasant accounts of events in the Upano—Palora region.

Finally, census records provided us with an unusual opportunity to corroborate casual observation with quantitative data. Unlike most Latin American governments Ecuador collects and publishes data on small, rural places, parishes within municipalities.[6] The existence of data on these places made it possible to carry out a comparative historical analysis of the advance of spontaneous colonization in adjacent valleys on the eastern slope of the Andes. These analyses provided statistical confirmation for the historical connections between lead institution activities and migratory trends.

To summarize, we searched through archives, analyzed satellite images, and conducted approximately one hundred and thirty-five interviews in our research on development and deforestation in Morona Santiago. This accounting of our efforts misses two vital aspects of the research process. First, participation in land clearing efforts long before we began this research left us with impressions that helped to shape our understanding of tropical deforestation. While personal experience frequently introduces biases that compromise the quality of research, it also can lead to a clear sense of the ideas that animate participants in an activity and, in so doing, contributes to an understanding of the overall process. Second, the field research also involved innumerable evenings spent in conversation with villagers about the history and current affairs of Morona Santiago. Observations and ideas generated through these conversations frequently provided the threads that stitched together the diverse pieces of information gathered through interviews, archival research, and remote sensing analyses. This mix of methods provided the empirical basis for our assertions about deforestation in southeastern Ecuador.

APPENDIX B
Shuar Landholdings by Region

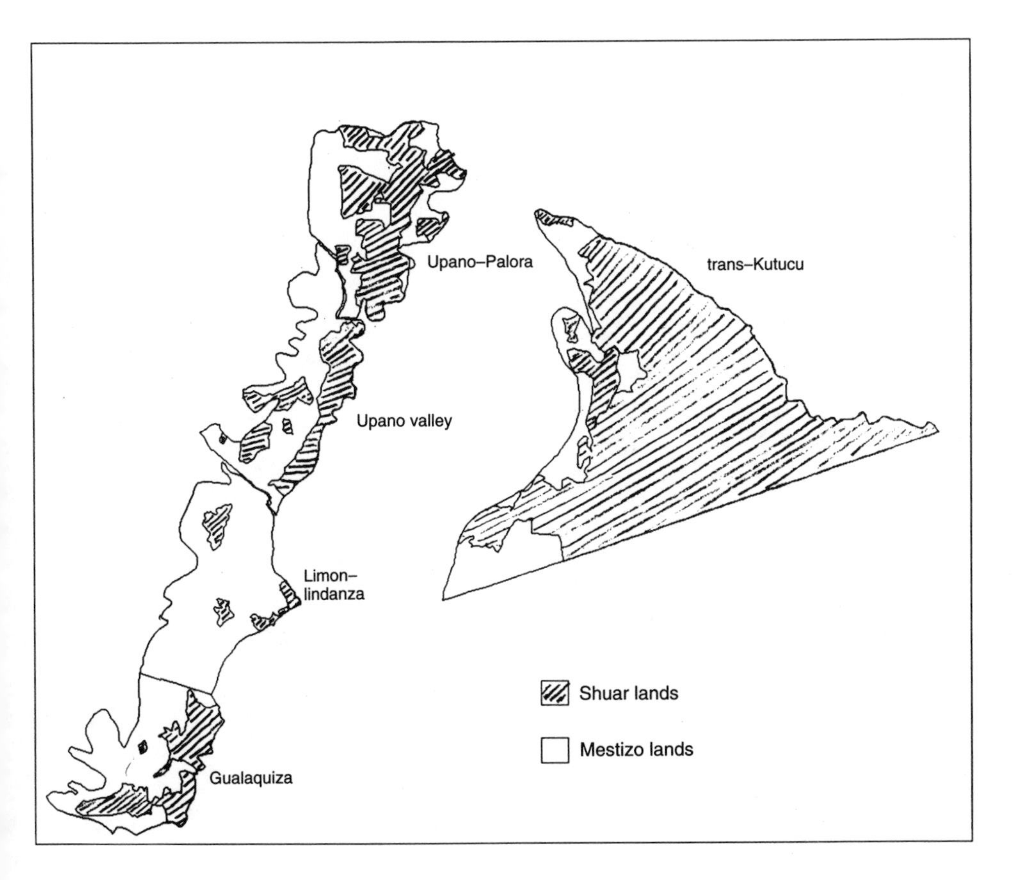

Notes

Introduction

1. The particular species making noise in the early mornings in the forests of northeastern Morona Santiago were, most likely, toucans (Ramphastos culminatus), parrots (Amazona farinosa), monkeys (S. sciureus), ocelots (Felis pardalis), and jaguars (Panthera onca).

1. The Geography of Tropical Deforestation

1. As with most definitions, this one has borderline cases. For example, it implies that conversion of a primary forest into a coffee grove would constitute deforestation, but the conversion of a forest into an agroforestry plantation would not. Calling one but not the other type of conversion "deforestation" exaggerates the difference between them. The two types of conversion differ in some, but not all, important respects. Their effects on biological diversity in a place might be quite different because the agroforestry plantation preserves more species, but their effects on local climates might be quite

similar because both land uses would maintain tree cover in a place. Because less than 5 percent of all tropical forests have been converted into forest plantations (FAO 1981a, 1981b), these borderline cases cover a relatively small area of the world's tropical forests.

2. A Theory of Tropical Deforestation

1. "Deforestation Slows." Science 251:1425, March 22, 1991.

2. This famous characterization of northeast Brazil and the Amazon basin comes from General E. Medici. He described the two regions in these terms in 1969 when he presented the Brazilian government's Transamazon road building and colonization scheme to the public.

3. This generalization is based on a global evaluation (Scudder 1981) and case studies in Asia (James 1983; Simkins and Wernstedt 1971; Lewis and Lewis 1971; Utomo 1967), Africa (Raison 1981:70–73), and Latin America (Painter and Partridge 1989:366; Eckstrom 1979:36; Casagrande, Thompson, and Young 1964).

4. There are several notable similarities and differences between growth coalitions in urban centers of developed countries and rain-forest frontiers in developing countries. First, the goals of coalition members in the two locales are similar. In both instances people join the coalitions as part of a strategy of economic accumulation. Second, the social and economic composition of the coalitions differs dramatically. In developed countries the coalitions bring together members of the local business elite, so they tend to be homogeneous and upper class in their composition. In the peripheral regions of developing countries the coalitions incorporate both poor peasants and individuals with substantial resources.

5. The Asian countries are Indonesia, Malaysia, the Philippines, and Thailand.

6. Beginning with Frederick Jackson Turner, students of the frontier have argued that frontiers have a "safety valve" effect. By providing economic opportunities for the poor, frontiers prevent surplus labor from accumulating elsewhere and, in so doing, prevent political discontent. Recent studies (Shresta 1989) have argued that modern frontiers do not have a safety valve effect because middle peasants and politically influential families end up owning a large proportion of the land on the frontier. The argument advanced here would agree with these analyses regarding the distribution of land on the frontier but disagree regarding the political effects of this pattern. If Wolf (1970) is correct in his argument that the crucial actors in most agrarian revolutions are middle peasants, then frontiers, by providing an array of economic opportunities for this strata of the peasantry, may actually dampen political discontent without redistributing resources to the poor.

7. Popkin (1979) argues that peasants quite frequently assume the free rider position in efforts to provide public goods. The accuracy of our characterization of peasants as free riders depends in part on the lead institution in

a rain-forest region. When oil and logging companies build roads, peasants are free riders, just as they are described in the text. When the government builds the roads, peasants may participate in a road's construction, either through the payment of taxes or through voluntary labor to repair or maintain the roads. In these later instances peasants are not free riders.

8. Marlise Simons, "In Amazon, Road Is a Dream, but to Ecologists a Nightmare." New York Times, 2/14/89.

3. The Ecuadorian Amazon: Land, People, and Institutions

1. Ecuador has 20,000–25,000 plant species and 2,436 animal species compared with 17,000 plant species for North America and 1,394 animal species for the continental United States (Cabarle et al. 1989:3–4, cited in World Bank 1990:5).

2. Prance (1973, 1976) has identified eastern Ecuador as the possible site of a Pleistocene refuge forest. Physical geographers theorize that eastern Ecuador remained forested when climactic changes caused a generalized reduction in the extent of tropical forests during the Pleistocene. The extraordinary age of the forests in the refugia should promote endemism, which in turn would lead to extraordinary levels of species diversity in places like eastern Ecuador. The field studies necessary to confirm these expectations have not been carried out.

3. In the early 1980s the Ministry of Agriculture, relying on data on colonization in the Amazon region and on the Coast, estimated tropical deforestation at 182,800 hectares per year. A study of tropical deforestation between 1977 and 1985, which used remote sensing techniques, estimated the loss of tropical forest at 340,000 hectares per year. Knowledgeable observers have estimated lower rates of deforestation, around 100,000 hectares per year. See Felipe Burbano, "La Amazonia: Un Territorio ya Repartido." Hoy (Quito), 9/2/85; "Alarmante Disminucion de Bosques Amazonicos." El Comercio (Quito), 3/12/87; World Resources Institute 1990:35–36.

4. For details about our previous work in Ecuador and its effect on our field research, please see appendix A.

5. Throughout the following analysis we will use Ecuadorian nomenclature to identify the major regions of Ecuador. People refer to the coastal lowlands along the Pacific as "the Coast," the Andean highlands as "the Sierra," and the Amazon lowlands as "the Oriente."

6. In 1989 the northern portions of the oil field region became a new province, Sucumbios.

7. The province of Morona Santiago was created with its present boundaries in 1954. Before 1954 Morona Santiago and Zamora Chinchipe were one province.

8. Rolando Tello Espinosa, "Morona, Tierra de Promesas Irrelizados." El Comercio (Quito), 9/4/89.

9. For centuries westerners have referred to these people as Jivaros, a term with derogatory implications in Spanish. They prefer to be called "Shuar," which means "the people" in their native tongue. "Shuar" has recently replaced "Jivaro" as their name in most reports of their activities, and we will follow this usage here. Anthropologists still use the term "Jivaroan" to describe the larger population of lowland indigenous peoples of whom the Shuar are one group.

10. F. Villaroel G., "San Juan Bosco, Pueblo Activo de Limon Indanza." El Comercio (Quito), 12/1/86.

11. Anonymous interviews, CREA colonists, Morona Santiago, 1971.

12. There are no census figures that separate out people by their ethnic identity. These figures come from CONAIE, a consortium of Amerindian organizations in Ecuador, and are estimates only. They are cited in Susana Cordero de Espinosa, "Nosotros, los Indigenas." Hoy (Quito), 5/20/89.

13. "Colonias Militares en el Oriente." El Comercio (Quito), 10/13/10.

14. Agustin Cueva, "Pionero de Estudios Agrarios Modernos: Pio Jaramillo Alvarado." La Liebre Illustrada (Quito), 1/08/89.

15. Instituto Nacional de Colonizacion, Plan Vial de Colonizacion Oriental, Quito, 1962.

16. Rafael Pesantez P., "Geopolitica y Integracion Amazonica." El Comercio (Quito), 7/19/86; Rafael Pesantez P., "Recuerdos de un Gran Orientalista." El Comercio (Quito), 10/31/89.

17. "Contrato de Colonizacion del Oriente." El Comercio (Quito), 9/24/10.

18. "Presidente Plaza Califica de Absurda la Colonizacion Masiva del Oriente." La Tierra (Quito), 2/23/50.

19. Gini coefficients, an index of land concentration, range from 0 (complete equality) to 1 (complete inequality), and average between 0.4 and 0.5 for most municipalities in the Amazon region. In the colonization zones on the coastal plain in northwestern Ecuador the Gini coefficients average around 0.6. Elsewhere on the Coast and in the Sierra the index of land concentration averages between 0.8 and 0.9 (Chiriboga, Landin, and Borja 1989:72–79).

20. "El CREA y el Oriente." El Mercurio (Cuenca), 2/13/72.

21. Marcia Mendoza A., "Hay que Volver Mirar al Campo." El Comercio (Quito), 8/23/86.

22. "Avanca Carretera Mendez—Morona." Hoy (Quito), 5/14/86.

23. Anonymous interviews, Morona colonists, Macas, 1989; "Colonos de Puerto Morona Piden Asistencia del IERAC." El Mercurio (Cuenca), 2/20/89.

4. Regional Development and Deforestation in Morona Santiago

1. Fernando Villaroel G., "En Villano Estan los Mejores Terrenos." El Comercio (Quito), 5/05/85.

2. Why Macas survived after 1599 when all other Spanish communities in the southern Oriente disappeared remains a mystery. One possible explana-

tion comes from a recently published study of the Incan origins of the Ecuadorian upper class (Noboa 1990). This book analyzes a will, which discusses Incan properties in and around Macas before the Spanish conquest (Noboa 1990:63). This evidence suggests that, despite the defeat of the Incan armies of Huayna Capac and Tupac Yupanki by the Shuar, the Incas were able to maintain a presence in Macas. This circumstance would explain how the Spanish got out to the Oriente so fast—they followed Incan trails. It would also explain why the Spanish were so successful at getting the Amerindians to extract gold for them—the Incas had already organized the extraction of gold. Finally, it would explain why the Spanish were able to survive at Macas but not elsewhere.

When the Shuar sacked and burned Sevilla de Oro on the east bank of the Upano, some of the Spanish escaped to an already fortified, formerly Incan position at Macas on the west bank of the Upano. This fallback position enabled the Spanish at Macas to survive the Shuar onslaught. In other parts of the southern Oriente the Spanish did not have such a refuge, so when the Shuar destroyed the Spanish cities, the cities' inhabitants either left the Oriente, or they died at the hands of the Shuar. Careful archeological work will be necessary before we can accept this interpretation of events as historically accurate.

3. "Los Misioneros." El Comercio (Quito), 12/10/11.

4. Fernando Villaroel G., "Indanza es un Pueblo de Colonizacion y Ganaderia." El Comercio (Quito), 11/24/86.

5. Correspondence between Salesian missionaries in Mendez and their superiors, 1917 and 1918, quoted in Bottasso 1982:154.

6. P. Jaramillo Alvarado, Informe al Gobierno, Direccion General del Oriente 923–1924, quoted in Bottasso 1982:154.

7. Fernando Villaroel G., "Copal no quiere Morir." El Comercio (Quito), 9/21/86.

8. Ibid.

9. Fernando Villaroel G., "San Juan Bosco, Pueblo Activo de Limon-Indanza." El Comercio (Quito), 12/1/86.

10. Fernando Villaroel G., "El Padre Jacinto Bucheli, un Misionero de la Selva." El Comercio (Quito), 5/26/85.

11. Interviews with ex-colonists, Cuchibamba valley, 1970, 1986.

12. Ibid.

13. Personal communication, P. Eckstrom, 1990.

14. Anonymous interview, former inspector of schools, Limon-Indanza, 1986

15. Anonymous interview, former inspector of schools, Limon-Indanza, 1986; anonymous interview, former San Carlos colonist, 1986.

16. Anonymous interview, former inspector of schools.

17. "Proyecto de Colonizacion Coangos—Cenepa." Departmento de Colonizacion, CREA. Cuenca, 1976.

18. Anonymous interview, former CREA promoter, Coangos project, 1986.

19. Anonymous interview, CREA extension agent and colonist in Pablo Sexio, 1989.

20. Villaroel, "San Juan Bosco ..."

21. L. Carollo, Observaciones sobre el libro Federacion de Centros Shuar, "Solucion Original a un Problema Actual," Sociedad Don Bosco, Cuenca, 1977.

22. J. Vigna, Memorandum para el Delegado del Ministerio del Gobierno, Mision Salesiana, 22 de Julio, 1947, reprinted in Bottasso 1982.

23. Federacion de Centros Shuar 1976; A. Sanchim,"Formacion y Historia del Centro Shuar Skanas"; A. Tsukanka, "Formacion y Historia del Centro Shuar Yukutais"; R. Wampash, "Formacion y Historia del Centro Shuar Chumpias"; all three manuscripts produced at Bomboiza: Instituto Normal Bilingue Intercultural Shuar, 1985.

24. Anonymous interview, former parish priest at the Salesian mission in Chiguaza, 1986.

25. Anonymous interview, former head of the Salesian mission in Sevilla Don Bosco, 1986.

26. For example, the population, all Shuar, in villages around Sevilla Don Bosco increased from 1,120 in 1950 to 2,598 in 1970. Because there was almost no off-farm employment in Sevilla Don Bosco, all of the additional households had to acquire land in order to subsist. Source: mission records, reprinted in C. Ochoa and L. Sierra 1976.

27. Anonymous interview, surveyor, Federacion Shuar 1980; C. Amaluiza and M. Segovia 1977; Leonardo Wisim, "Formacion y Historia del Centro Shuar Tsuirium." Bomboiza: Instituto Normal Bilingue Intercultural Shuar 1985.

28. Federacion de Centros Shuar, "Los Shuar: Colonizadores del Oriente." Sucua, 1973.

29. A. Abad, "Informe Preliminar, Projecto de Colonizacion," San Jose de Morona, Departmento de Colonizacion, CREA 1975.

30. Chicham, Publicacion de la Federacion Shuar, January 1981.

31. "Director del CREA mantuvo reunion con Federacion Shuar." El Mercurio (Cuenca), 9/24/79.

32. Chicham, January 1981.

33. "Atestigan Entrega de Tierras a Shuaras." Hoy (Quito), 4/3/90.

34. The boundaries of each subregion and the location of Shuar lands in each subregion are indicated in the map in appendix B. To estimate the extent of the Shuar lands, we used remote sensing images to establish the boundaries of the lowlands in a subregion. We limited our subsequent calculations to the lowlands in the belief that neither the colonists nor the Shuar wanted control over the mountainous uplands in the region. In other words the two groups compete only for lands at lower elevations. With the boundaries of the lowlands in each subregion established, we digitized the resulting polygons using

GRASS, a geographic information system. We then calculated the proportion of each polygon occupied by Shuar centros. The boundaries of the centros came from maps produced by the Federacion de Centros Shuar.

35. A Sanchim, "Formacion y Historia del Centro Shuar Skanas."

36. Federacion de Centros Shuar 1976:125; Informe, Asamblea General, Vicariato Apostolico de Mendez, March 1985.

37. Anonymous interviews, topographers, Federacion de Centros Shuar, Sucua, 1986.

38. Anonymous interviews, former directors, Federacion de Centros Shuar, 1989, 1990.

39. Anonymous interview, former parish priest at the Salesian mission in Chiguaza, 1989.

40. We computed the statistics on loan recipients from data in Federacion de Centros Shuar, Solucion Original a una Problema Actual (1976:270–276).

41. To control in this analysis for the contaminating effects of a location near a growth pole like Taisha, we only compared centros that were more than twenty kilometers from Taisha or Macuma, the two mission communities east of the Kutucu. For this reason we refer to these communities as "interior" villages. We performed the remote sensing analyses using a 1983 multispectral scanner image. A 1985 map from the federation indicated the boundaries of the villages. For further details, see appendix A.

42. "Hay que Presionar para una Mejor Atencion." El Mercurio (Cuenca), 6/25/89.

43. "Los Achuar Declaran Defensa de Territorio." America Indigena, July 1988.

44. "Gestionan Soluciones para Problems de Grupos Shuar." El Comercio (Quito), 8/6/85.

45. Remote sensing data provided the crucial data on the extent and location of cleared land in the 1980s. Information about the date of the first land clearing in a location came from a variety of sources. For the early period the records and recollections of the Salesian missionaries proved important. For the later period, post-1968, personal recollections provided some information. Interviews with a wide range of informants provided information about deforestation in places that we could not visit.

46. Variations in cloud cover and topography account for the different numbers of subareas analyzed in each of the regions.

47. Because the multispectral scanner classifies data in 80-meter chunks it misses small clumps of trees in a pasture or around a house, and it misses small gardens in a forest. In effect it overestimates the extent of the dominant land use in an area. Presumably the effects of these errors cancel out in analyses of a large area.

48. The only survey of land use among Shuar landholders in the region, done in one centro, also indicates clearings that total about 5 percent of the land (Bernal and Torres 1978:120).

5. Trailblazing in a Large Forest: The Upano—Palora Plain in the 1960s

1. Anonymous interview, Salesian missionary, Cuenca, 1986.

2. For additional details on this pattern of land clearing and the associated conflicts over land, see Rudel 1983.

3. Anonymous interview, former Peace Corps—CREA promoter, 1989. A number of analysts have contended that a thorough land reform would reduce migration to rain forest regions by creating new economic opportunities in the sending regions. The situation in the province of Canar during the late 1960s supports this contention. IERAC had decided to expropriate several haciendas, and the prospect of land in the Sierra reduced the peasants' interest in colonization in the Oriente. George Mowry, "Colonizacion en Canar." Programa de Colonizacion, Cuerpo de Paz, 1969.

4. Departmento de Colonizacion, "Estudio de las Condiciones de los Campesinos que se Integran a Programas de Colonizacion." CREA, Cuenca, 1975.

5. During the first year after settlement about 10 percent of the group got lost and spent an unanticipated night in the forest. The fear of snakes had a real basis. Every week during the first year the colonists would encounter venomous bushmasters (Lachesis muta) whose bite could in some circumstances kill a person. After the first year the number of encounters with poisonous snakes declined sharply. In all probability the colonists had killed a large proportion of the bushmasters in the region. Ocelots (Felis pardalis) and jaguars (F. onca) also inhabited the area initially, but they quickly moved out when human activity increased, so the colonists rarely encountered and did not fear the cats.

6. Villaroel, "Indanza es un Pueblo de Colonizacion y Ganaderia."

7. The colonists in Sinai, as elsewhere, made heavy use of pack animals to transport their provisions, but conditions in these frontier settings wore out the horses and mules. In the first year of settlement before pasture grasses had matured in the new town, the colonists had no forage to feed their horses and mules when they arrived in Sinai, exhausted after carrying heavy loads over rough trails. Sometimes they would nibble at salted palm leaves the colonists would put out for them. The pack animals did not get any pasture grass to eat until the following day when they arrived back at their original point of departure with its abundant pastures. Two months of this regimen weakened even the strongest pack animals to the point where they could no longer be used. During the first year of Sinai's existence, the colonists ruined about fifteen mules and horses in this fashion. With the pack animals too weak to work, people had to carry the foodstuffs into the new town on their backs.

8. Tim Lacy and Tom Rudel, "Sinai—The First Two Months." Colonization Program, U.S. Peace Corps, 1969.

9. Anonymous interview, former evaluator, CREA colonization program, 1986.

10. "San Carlos de Zamora," informe de Teofilo Iniguez, Representante de la Cooperative, Departmento de Colonizacion, CREA 1969.

11. Cited in E. E. Hegen 1966:43.

12. José Gavilanes, "Ayer, Hoy, y Manana." El Ecuador (1970), 2(3):38–39; anonymous interview, Sinai, August 1986.

13. In the mid-1970s CREA decided not to build the section of the Macas—Puyo road between the Upano and the Palora rivers, presumably because the new road would open up access to markets in the northern Sierra, such as Quito and Ambato. Merchants from these places would then challenge the economic dominance of Cuenca's merchants in the region (Rudel 1989b). Once CREA declined to build the road, influential families from Macas were able to get the Ministry of Public Works to change the route of the new road so that it crossed the Upano at Macas. This change in route did not affect the fortunes of the CREA colonists or the large landowners because the road still passed close to their lands, but it did lead to a series of intractable problems building the bridge. By 1990 three bridges had been built and collapsed at the Macas crossing of the Upano.

14. Anonymous interviews, Sinai colonists, 1989.

15. "Shuaras Atacan a Colonos." Hoy (Quito), 9/20/90; "Colono fue quien invadio terrenos." Hoy (Quito), 9/21/90.

16. Gonzalez and Morocho 1979; "Derechos Shuar son Violados." El Mercurio (Cuenca), 5/17/86; Departmento de Colonizacion, Informe sobre el Proyecto de Colonizacion San Jose de Morona: los Primeros Diez Meses, CREA, Cuenca, 1976.

17. Anonymous interviews, Macas, 1986.

6. Clearing Forest Remnants: The Upano—Palora Plain in the 1980s

1. Censo de Cultivos—Sinai, Departmento de Colonizacion, CREA, Marzo, 1978.

2. Map: Ubicacion de Centros Shuar en Morona Santiago, Federacion de Centros Shuar, 1985. Two minor sources of error in delineating community boundaries are worth noting. First, the map outlines the boundaries of Shuar communities and adjoining colonist communities, but it does not indicate boundaries between adjacent colonist communities. In these instances we have chosen blocks of land, centered on the parish's urban center, to indicate the areal extent of the parish. The true boundaries of the parishes deviate some from these boundaries. Second, because the subset program in ERDAS, the software package used in the remote sensing analyses, only subsets blocks of land, it was impossible to include all of the land in irregularly shaped communities in each analysis. In some instances where a community had an irregular shape, we analyzed a series of blocks of land which together approximated the irregular shape of the community's land.

3. The image under analysis is a multispectral scanner image with a minimum resolution of 80 meters square. For further details regarding the analysis, see appendix A.

4. The ethnic identity of a community is always clear because residential segregation between colonists and Shuar is close to 100 percent. Colonists never live in centros because they cannot own land in them. While the Shuar can be found living outside of centros in the Upano valley, their numbers are small and declining (Amaluiza and Segovia 1977).

5. In this analysis large landholdings contain more than 500 hectares of land.

6. The regression analysis of deforestation, reported in table 6.1, has been weighted by the land area of the community. This procedure prevents communities covering small areas from exercising an unwarranted influence in the analysis.

7. When a household owned several tracts of land, the tracts were treated as a single unit in the analysis.

8. We did not interview landowners who settled in the outlying areas of Sinai, beyond the Ambusha River, a one- to three-hour walk from Sinai, because this area was settled somewhat later than the lands near the parish center and for that reason would not be fully comparable with tracts of land in Uunt Chiwias. For more detail about the design of the study, sampling procedures, and the questions in the interview schedule, see appendix A.

9. Very little reforestation has occurred in the zone, so the proportion of land that has been cleared and the proportion of land currently being used for agriculture are the same for almost all of the units.

10. Censo de Ganado—Sinai, Departmento de Colonizacion, CREA, Agosto, 1973.

11. Anonymous interview, former CREA—Peace Corps extension worker, 1990.

12. Although most colonists did not have official titles to the land, they received provisional titles during the 1970s that made it possible to use their land as collateral in applying for a loan.

13. Agusto Abad, "Informe: Cooperativa Sinai," Departmento de Colonizacion, CREA, June 1973.

14. This pattern underscores the importance of family composition, as emphasized by Chayanov (1966), in understanding the evolution of family farms.

15. Anonymous interview, Salesian missionary, Chiguaza, 1986.

16. Anonymous interviews, former directors, Federacion de Centros Shuar, 1990.

17. Federacion de Centros Shuar 1976:270–276.

18. The politically weak centros are Pajanak and Tsemantsmaim; the two centros with reputations for strong leadership are Mutintsa and Kuchaents. Because it is difficult to measure the strength of political leadership in many

centros, a test of this hypothesis on a large sample of centros would require an extensive effort at data collection which we did not undertake.

19. Tomas Pujupat, "Formacion y Historia del Centro Shuar Pumpuis," Instituto Nacional Bilingue Intercultural Shuar, Bomboiza, 1985; anonymous interview, CREA—Peace Corps extension worker, 1989.

20. Technically, profits through the sale of land should not occur among the Shuar because the sale of individual tracts of land within centros is prohibited as one of the stipulations in the global title to land which each village holds. Nonetheless, individuals do on occasion sell their land to other Shuar and presumably earn profits from the creation of pastures on the land and the construction of roads or airports nearby. The illegal nature of the sales makes it difficult to estimate the frequency with which it occurs. Anonymous interviews, former directors, Federacion de Centros Shuar, 1990.

21. Personal communication, J. W. Hendricks, 1990.

22. Anonymous interviews, colonists, Sinai, 1986; anonymous interviews, Shuar, Sucua, 1986.

23. Data sources: for Sinai during the 1970s, Agusto Abad, "Informe: Cooperativa Sinai"; for the Shuar during the 1970s, Luis Ortiz Vasquez, "La Poblacion Shuara en la Zona Chiguaza: Analysis Socio-economico." CREA 1972; Federacion de Centros Shuar 1976:69. The data for 1986 came from our surveys in Sinai and Uunt Chiwias

24. F. Villaroel G., "Un dia con los Shuar de Wapu." El Comercio (Quito), 9/07/86.

25. Censo de Cultivos—Sinai, Departmento de Colonizacion, CREA, Marzo, 1978; survey by authors, 1986.

26. Anonymous interview, extension agent, CREA—Pablo Sexto, 1989.

27. Personal communication, Bradley Bennett, 1991.

28. Anonymous interview, forester, German volunteer corps, Macas, 1989.

7. The Second Generation's Search for New Lands

1. Anonymous interview, extension agent, CREA—Pablo Sexto, 1989.

2. Anonymous interviews, Sinai residents, 1986, 1989.

3. Differences in stocking rates between large and small landowners could account for these differences. A study in the southern portions of the Upano valley found that smallholders stock their pastures at higher rates than large landowners (Santana 1989). We found a similar pattern in Sinai and Uunt Chiwias.

4. It is possible that the pastures are almost always green because they are new and experience rapid regrowth. The small amount of new pasture in Uunt Chiwias casts some doubt on this interpretation of the data.

5. To correct for the skewed distribution of denuded pasture across communities, we logged the variable; then we divided it by the amount of pasture in the community. Places with a lot of healthy pasture scored low on the

resulting measure; places with little healthy pasture scored high on the measure. This variable was then regressed on the three independent variables.

6. As we did in the regression analysis of deforestation, we weighted the regression analysis of pasture degradation by the amount of pasture in the community. Using this procedure, communities with little pasture weigh less in the analysis than communities having more pasture.

7. Anonymous interview, cattle specialist, U.S. Peace Corps, Macas, 1989.

8. During the 1974–1982 period Morona Santiago had the lowest interprovincial net migration rate, +3.1 percent, of the four Amazon provinces (CONADE 1987:187).

9. As in the previous chapter, households are the units of analysis. There are potential problems of selection bias with this sample. The participants in the Ebenecer episode illustrate the problem. If the most successful of those people who search for new lands leave Sinai when they find new land as the Solis and Guzman families did when they founded Ebenecer, then a sample of Sinai residents eight years later will not include them or other families that have moved to the new lands. In this sense a cross-section of Sinai residents would underestimate the number of movers and produce biased estimates of their characteristics. While the nature of the problem is clear, its magnitude is relatively small. We could find no other families who had left Sinai during the 1978—1989 period for new lands. Almost all of those families that acquired new lands during the period retained a house in Sinai and are included in our sample. Because the Uunt Chiwias Shuar were less likely to move than Sinai's colonists during this period, the magnitude of this problem in the Shuar subsample should also be small.

10. These estimates would, for example, reflect the surge of income that occurs when a household head sells assets to meet medical expenses.

11. The logistic regression analyses presented in table 7.2 produce maximum likelihood estimators with a goodness of fit statistic (chi square) rather than a r2 as in conventional regression analyses. The coefficients and their standard errors are comparable to the coefficients and standard errors in regression analyses.

12. In a comparative study of Sinai and an unnamed Shuar centro in 1977, Navas and Lara (1978) found a similar difference between colonist and Shuar incomes. The average annual income for the colonists was US$624; the comparable figure for the Shuar was US$248. Navas and Lara do not indicate how they measured income, so these figures are open to question.

13. Anonymous interviews, surveyors, Departmento de Topographia, Federacion de Centros Shuar, Sucua, 1986; anonymous interviews, former directors, Federacion de Centros Shuar, Quito, 1990; anonymous interview, Tsemantsmaim Shuar, 1986.

14. Anonymous interviews, surveyors, Federacion de Centros Shuar; Pamela E. Isreal, "Interim Report of Field Research Among the Shuar," Fulbright Commission: Ecuador, 1984.

15. The snow-capped peaks visible from Sinai are Sangay, El Altar, and Tungurahua.

16. Anonymous interview, regional chief, Parque Nacional Sangay, San Isidro, 1986.

17. E. J. Ulloa, "El Parque Nacional Sangay." El Espectador (Riobamba), 8/12/85.

18. Anonymous interview, regional chief, Parque Nacional Sangay; Anne Macey, Gordon Armstrong, Menard L. Hall, and Nelson Gallo, "Sangay: Estudio de las Alternativas de Manejo," Departmento de Parques Nacional, Direccion Generale de Desarrollo Forestal, 1976; "Parque Nacional Sangay: Plan de Manejo," Departmento de Areas Naturales y Vida Silvestre, Programa Nacionale Forestale, 1982.

19. Anonymous interview, regional chief, Parque Nacional Sangay.

20. Anonymous interview, former CREA—Peace Corps extension worker, 1990.

21. The surveys with farmers in Sinai indicate that 76 percent of their land has been cleared while the remote sensing analysis indicates that only 55 percent of the land has been deforested. The discrepancy between the two methods of calculating deforestation occurs because the remote sensing analysis includes all land, including areas in deep ravines, most of which remains public land and all of which is forested. The extent of deforestation calculated from the interviews with farmers does not include public lands, so this method yields a higher estimate of the deforested area. A similar line of reasoning would explain most of the discrepancy between the extent of deforestation across the Ambusha as measured during the Yaccino survey and the extent of deforestation indicated by the remote sensing analysis.

22. Census of households, Sinai, 1986.

23. "Evaluacion de la Colonizacion y Analisis de la Poblacion Indigena en la Provincia de Morona Santiago." CREA, Cuenca, 1976.

24. Sources for La Quinta and La Sexta region, anonymous interviews with colonists from these regions; for La Septima and La Octava cooperatives, anonymous interview, CREA extension worker Pablo Sexto, 1989.

25. Anonymous interviews, former directors, Federacion de Centros Shuar, Quito, 1990; anonymous interviews, topographers, Federacion de Centros Shuar; F. Riofrio, "Franceses y Shuaras Frente a Frente: Un Apreton de Manos." El Comercio (Quito), 12/13/89.

26. Anonymous interviews, topographers, Federacion de Centros Shuar, 1986; Fernando Villaroel G., "Las Fronteras Vivas en el Valle del Nangaritza." El Comercio (Quito), 6/10/86.

27. Anonymous interview, regional chief, Parque Nacional Sangay.

28. ersonal communication, Peter Eckstrom, 1990.

8. Tropical Deforestation: An Assessment with Policy Implications

1. Anonymous interview, IERAC administrator, Lago Agrio, 1971.

2. Field notes, Lago Agrio office of IERAC, 1971.
3. Anonymous interview, IERAC topographer, Lago Agrio, 1971.
4. These estimates of deforestation come from aerial photographs of the region, taken in 1978, and analyzed by Hiroaka and Yamamoto (1980:431).
5. Anonymous interviews, colonists, Cooperative Nueva Loja, Lago Agrio, 1971.
6. The same pattern characterizes landholdings on the coast. A survey of colonists' farms near Santo Domingo on the coast found 46 percent of the land on small farms in forest, 49 percent of the land on middle-sized farms in forest, and 60 percent of the land on large farms in forest. Small farms averaged 14.1 hectares; middle-sized farms averaged 29.8 hectares, and large farms averaged 54.9 hectares of land (Gladhart 1972).
7. F. Villaroel G., "La Fiesta Nativa," El Comercio (Quito), 3/7/88.
8. James Brooke, "New Effort Would Test Possible Coexistence of Oil and Rain Forest." New York Times, 2/26/91.
9. Marlise Simons, "The Smelters' Price: A Jungle Reduced to Ashes." New York Times, 5/28/87.
10. "Piden Fijar Territorios en Amazonia." Hoy (Quito), 10/15/89; James Brooke, "Conflicting Pressures Shape the Future of Brazil Indians." New York Times, 2/25/90.
11. Von Thunen (1851) would have predicted these problems. According to his theory, farmers can only grow a diverse set of products profitably if their farms are close to large markets. The higher marketing costs that must be borne by farmers farther from the market limits their choices regarding crop mix and compels them to focus on the one or two products they can market profitably (Dunn 1954). Farmers outside of a market economy, like some of the peasants practicing agroforestry in Peru, will of course grow a diverse set of products for household consumption.
12. For a comprehensive discussion of social forestry programs in Southeast Asia, see Poffenberger (1990).
13. "A Future for the Tropical Forest Action Plan." Forest Conservation Programme Newsletter. International Union for the Conservation of Nature, no. 12, January 1992.
14. Julian Caldecott, "Biodiversity Conservation Outside Protected Areas." Forest Conservation Programme Newsletter. International Union for the Conservation of Nature, no. 11, June 1991.

Appendix A: Reseaarch Methods

1. Neither one of us was ever refused an interview. The ready acceptance of both us and our project probably stems from our extensive early experience with both groups of respondents.
2. Although the Shuar are not technically the "owners" of any land which they work within a centro, they can pass centro land on to their children, and

people generally refer to a particular Shuar in a centro as the "owner" of this or that tract of land.

3. Women as well as men own land in Sinai and Uunt Chiwias, so some of our respondents were women.

4. When we began the study, SPOT data from French satellites was not available, so we had to use either TM or MSS data gathered by LANDSAT satellites.

5. The chief difficulty in doing the image differencing analysis involved the programming necessary to get an early 1973 MSS tape of the same region into a format compatible with the 1983 MSS tape.

6. A parish usually covers an area and a population equivalent to those places and populations served by a post office in rural regions of the United States.

Glossary

Environmental Social Science Terms

Agroforestry: agriculture that emphasizes the cultivation of tree crops.

Colonization: the occupation and settlement of a sparsely populated area by farmers.

Colono system: a once common tenure arrangement in Latin American frontier regions in which peasants and small farmers clear land along the fringes of the forest, farm it for several years, and then give it, voluntarily or involuntarily, to large farmers who put the land into pasture. The peasants then move farther into the forest to start a new farm and begin the cycle of deforestation, cultivation, and dispossession over again.

Growth coalition: a group of individuals who acquire land in a place and work together to develop it. They hope to profit from the increases in land prices that occur when people intensify land use in a place.

Immiserization: long-term economic decline among already poor people which makes it difficult for them to satisfy basic needs for food, shelter, clothing, and health care.

Integrated rural development: a set of complementary government programs aimed at stimulating economic activity and improving human welfare in a particular valley, river basin, or province.

Lead institutions: well-capitalized institutions that build the infrastructure, usually roads, which open up rain-forest regions for settlement and deforestation.

Penetration road: the first road to be built into a rain—forest region.

Pioneer front: an area along the margins of a large forest where farmers carve farms out of the forest next to one another and over time move in unison into a large block of forest, clearing land as they go.

Political ecology: an analytic approach in environmental social science which emphasizes the political economic interests of natural resource users.

Proletarianization: the process through which landowning peasants and small farmers lose their land to wealthy agrarian or industrial elites.

Slash mulch polyculture: a form of shifting cultivation, practiced in extremely humid places, in which cultivators leave felled trees and stumps to rot instead of burning them.

Smallholder: a peasant or farmer who cultivates a small tract of land.

Social forestry: a set of programs that encourages peasants to reforest the lands they control. These programs also promote common property ownership of forested lands which in previous years were open to exploitation by anyone.

Tropical deforestation: in our usage tropical deforestation occurs whenever people clear more than 40 percent of the trees in a primary tropical forest with a closed canopy.

Spanish/Quichua Terms

A medias: a contractual arrangement in which a small farmer receives every other offspring from a herd of cattle in return for pasturing the cattle on his lands.

Barbecho: the secondary forest and vegetation that returns after someone clears the primary forest, cultivates the land, and then abandons it.

Cana guadua: bamboo trees.

Centro: a group of Shuar who live in one locale and jointly hold or claim title to land in that area.

Colono: a peasant who moves to a new location to start a farm in an unclaimed tract of forest.

Cordillera: mountain range.

Empalisada: a trail covered by an unending row of logs laid parallel to one another; by stepping on the logs pedestrians do not become mired in the mud along the trail, so they can make much better time in walking from village to village than they could without the log surface.

Hacienda: a large landholding characterized by extensive agriculture.

Hacendado: the owner of a large hacienda.

Huasipungo/aje: in this arrangement an Ecuadorian peasant received land in return for working a certain number of days each week on a landlord's fields. Until the 1964 land reform abolished it, the huasipungaje limited a peasant's mobility by tying him to a particular hacienda.

Latifundia: large farm.

Mestizo: someone of mixed Spanish and Amerindian descent.

Minifundia: small farm.

Nudo: a knot; a low transverse ridge in the Ecuadorian Andes which connects the two main north – south mountain ranges to one another.

Occidental: western.

Oriental: eastern.

Oriente: Ecuador's eastern lowlands, drained by left bank tributaries of the Amazon.

Respaldo: a row of farms that usually runs parallel to a transportation route, either a river or a road.

Sierra: Ecuador's Andean highlands.

Sindico: The resident of a Shuar centro who officially serves as its leader.

References

Adams, R. M. 1988. "Foreword." In J. Denslow and C. Padoch, eds., Peoples of the Tropical Rain Forest. Berkeley: University of California Press.

Adas, Michael. 1983. "Colonization, Commercial Agriculture, and the Destruction of the Deltaic Rainforests of British Burma in the Late Nineteenth Century." In R. Tucker and P. Richards, eds., Global Deforestation and the Nineteenth-Century World Economy, pp. 95–110. Durham, N.C.: Duke University Press.

Aguirre Beltran, Gonzalo. 1979. Regions of Refuge. Monograph #12, Society for Applied Anthropology. Washington, D.C.

Allen, Elizabeth. 1975. "New Settlement in the Upper Amazon Basin." Bulletin of the Bank of London and South America. 9:622–628.

Allioni, Miguel. 1910 (1978). La Vida del Pueblo Shuar. Sucua: Mundo Shuar.

ALOP-CESA-CONADE-FAO-MAG-SEDRI. 1984. La Situacion de los Campesinos en Ocho Zonas del Ecuador. Quito.

Aluma, John, C. Drennon, J. Kigula, S. Lawry, E. Muwanga-Zake, and J. Were. 1989. "Settlement in Forest Reserves, Game Reserves, and National Parks in Uganda: A Study of Social, Economic, and Tenure Factors Affecting Land Use and Deforestation in Mabira Forest Reserve, Kibale Forest Reserve, and Kibale Game Reserve/Corridor." Land Tenure Center Research Paper no. 98. University of Wisconsin, Madison.

Amaluiza, C. and M. Segovia. 1977. Un Grupo Shuar Marginado y Dependiente. Sucua: Mundo Shuar.

Ampudia, Telmo Carrera. 1987. Historia de la Tierra de los Macas. Macas: Consejo Provincial de Morona Santiago.

Anderson, A. B., ed. 1990. Alternatives to Deforestation: Steps toward Sustainable Use of the Amazon Rain Forest. New York: Columbia University Press.

Aramburu, Carlos E. 1984. "Expansion of the Agrarian and Demographic frontier in the Peruvian selva." In Marianne Schmink and Charles Wood, eds., Frontier Expansion in Amazonia, pp. 153–179. Gainesville: University of Florida Press.

Armstrong, Warwick and T. McGee. 1985. Theatres of Accumulation: Studies in Asian and Latin American Urbanization. London and New York: Metheun.

Barral, Henri. 1979. Poblamiento y Colonizacion en la Provincia de Esmeraldas y Comparison con la Zona de Colonizacion del Nororiente. Programa Nacional de Regionalizacion Agraria. Quito: Ministerio de Agricultura.

Barral, H. and C. Orrego. 1978. Informe sobre la Colonizacion en la Provincia del Napo y las Transformaciones en las Sociedades Indigenas. Mag-Orstrom, Quito.

Barral, H., R. Oldeman, and M. Sourdet. 1976. Reflexiones acerca del Estado Actual y del Porvenir de la Colonizacion del Nororiente. MAG-Orstrom, Quito.

Barsky, Osvaldo. 1984. La Reforma Agraria Ecuatoriana. Quito: Corporacion Editora Nacional.

Becker, Bertha. 1986. "Spontaneous/induced rural settlements in Brazilian Amazonia." In United Nations Center for Human Settlements, Spontaneous Settlement Formation in Rural Areas. Vol. 2: Case Studies, pp. 121–145. Nairobi.

Bernal, Eugenio and Hugo Torres. 1978. "La Zona de Morona: Proyecto de Colonizacion Ejecutado por el CREA." Thesis, Facultad de Ciencas Economicas, Universidad Estatal de Cuenca, Cuenca.

Bilsborrow, R. E. 1991. "Demographic Processes, Rural Development, and Environmental Degradation in Latin America." Paper presented at the meetings of the Latin American Studies Association, Washington, D.C.

Blaikie, Piers and Harold Brookfield. 1987. Land Degradation and Society. London and New York: Metheun.

Boster, James. 1983. "A Comparison of the Diversity of Jivaroan Gardens with that of the Tropical Forest." Human Ecology 11(1):47–68.

Bottasso, J. 1982. Los Shuar y Las Misiones: Entre La Hostilidad y El Dialogo. Quito: Ediciones Mundo Shuar.

Bowman, Isaiah. 1931. The Pioneer Fringe. New York: American Geographical Society, Special Publication #13.

Branford, Sue and Oriel Glock. 1985. The Last Frontier: Fighting Over Land in the Amazon. London: Zed Books.

Brito, E. 1938. Homenaje del Ecuador a Don Bosco Santo. 3 vols. Quito: Editorial Don Bosco.

Browder, John. 1988. "Public Policy and Deforestation in the Brazilian Amazon." In R. Repetto and M. Gillis, eds., Public Policies and the Misuse of Forest Resources, pp. 247–297. Cambridge, Eng.: Cambridge University Press.

Browder, J. and B. Godfrey. 1990. "Frontier Urbanization in the Brazilian Amazon: A Theoretical Framework for Urban Transition." Working Paper 90-1. Center for Urban and Regional Studies, Virginia Polytechnic Institute, Blacksburg, Va.

Brown, Lawrence A., R. Sierra, and D. Southgate. 1992. "Complementary Perspectives as a Means of Understanding Regional Change: Frontier Settlement in the Ecuadorian Amazon." Environment and Planning A, 24:1939–1961.

Brownrigg, Leslie. 1981. "Economic and Ecological Strategies of Lojano Migrants to El Oro." In N. E. Whitten, Jr., ed., Cultural Transformations and Ethnicity in Modern Ecuador, pp. 303–327. Urbana: University of Illinois Press.

Bylund, E. 1960. "Theoretical Considerations Regarding the Distribution of Settlement in Inner North Sweden." Geografiska Annaler B 42:225–231.

Cabarle, B. J. 1988. "An Assessment of Biodiversity and Tropical Forests for Ecuador." Paper prepared for the United States Agency for International Development, Quito.

Caldwell, J. 1990. "Will African Fertility Fall? The Nature of a Unique Society." Contemporary Sociology 19(3):410–412.

Campbell, D. T. 1988. Methodology and Epistemology in the Social Sciences. Chicago: University of Chicago Press.

Carrion, Lucia and M. Cuvi. 1985. La Palma Africana en el Ecuador: Technologia y Expansion Empresarial. Quito: Facultad Latinoamericana de Ciencias Sociales.

Casagrande, Joseph, Stephen Thompson, and Philip D. Young. 1964, "Colonization as a Research Frontier: The Ecuadorian Case." In Robert A. Manners, ed., Process and Pattern in Culture, pp. 281–325. Chicago: Aldine.

Castro, A. P. 1988. "Southern Mount Kenya and Colonial Forest Conflicts." In John F. Richards and Richard P. Tucker, eds., World Deforestation in the Twentieth Century, pp. 37–54. Durham: Duke University Press.

CEDIG (Centro Ecuatoriano de Investigacion Geografica). 1985. Poblaciones de las Parroquias, Ecuador, 1950–1982. Documentos de Investigacion. Serie Demografia y Geografia de la Poblacion, No. 2. Quito.

Chayanov, Alexander V. 1966. "Peasant Farm Organization." In D. Thorner, B. Kerblay, and R. E. F. Smith, eds., The Theory of Peasant Economy, chs. 2 and 3. Homewood, Ill.: R. D. Irwin.

Chiriboga, M., R. Landin, and J. Borja. 1989. Los Cimientos de una Nueva Sociedad: Campesinos, Cantones, y Desarrollo. Quito: Instituto Interamericano de Cooperacion para la Agricultura.

Christ, Raymond E. and C. M. Nissly. 1973. East from the Andes: Pioneer Settlements in the South American Heartland. Gainesville: University of Florida Press.

CIDA (Comite Interamericano de Desarrollo Agricola), 1965, Tenencia de la Tierra y Desarrollo Socio-Economico del Sector Agricola: Ecuador. Washington, D.C.: Union Panamericana.

Collins, Jane. 1986. "Smallholder Settlement of Tropical South America: The Social Causes of Ecological Destruction." Human Organization 45(1):1–10.

——. 1988. Unseasonal Migrations: The Effects of Rural Labor Scarcity in Peru. Princeton, N.J.: Princeton University Press.

CONADE (Consejo Nacional de Desarrollo). 1987. Poblacion y Cambios Sociales: Diagnostico Sociodemografico del Ecuador, 1950–82. Quito: Corporacion Editora Nacional.

Connell, J., B. Dasgupta, B. Laishley, and M. Lipton. 1976. Migration from Rural Areas: The Evidence from Village Studies. Delhi: Oxford University Press.

CREA (Centro de Reconversion Economica de Azuay, Canar, y Morona Santiago). 1975. Estudio de las Condiciones Socio-Economicas de los Campesinos que Se Integran a Programas de Colonizacion. Cuenca.

Crespo, Teodoro. 1961. El Problema de la Tierra en el Ecuador. Quito: Editorial Casa de la Cultura.

Dembner, S. 1991. "Provisional Data from the Forest Resources Assessment Project." Unasylva. 164:40–45.

Denslow, J. S. and C. Padoch, eds. 1988. People of the Tropical Rain Forest. Berkeley: University of California Press.

Descola, Philippe. 1981. "From Scattered to Nucleated Settlement: A Process of Socioeconomic Change Among the Achuar." In N. Whitten, ed., Cultural Transformations and Ethnicity in Modern Ecuador, pp. 614–646. Urbana: University of Illinois Press.

Domike, A. L. 1970. "Colonization as an Alternative to Land Reform." Analytic Paper #6. Washington, D.C.: United States Agency for International Development.

Dourojeanni, Marc. 1979. Desarrollo Rural Integral en la Amazonia Peruana con Especial Referencia a las Actividades Forestales. Lima: Departamento de Bosques, Universidad Nacional Agraria.

Dunn, E. S. 1954. The Location of Agricultural Production. Gainesville: University of Florida Press.

Eckstrom, John Peter. 1979. "Colonization East of the Andes: Responding to a New Ecology." Ph.D. dissertation, University of Illinois.

Eden, M. J. 1990. Ecology and Land Management in Amazonia. London and New York: Belhaven Press.

FAO (Food and Agriculture Organization: United Nations Environmental Program). 1981a, Tropical Forest Resources Assessment Project: Forest Resources of Tropical Africa, Regional Synthesis. Rome.

——. 1981b, Tropical Forest Resources Assessment Project: Forest Resources of Tropical Asia, Country Profiles. Rome.

Fearnside, Philip. 1989. "Deforestation in Brazilian Amazonia: The Rates and Causes of Forest Destruction." The Ecologist 19(6):214–218.

——. 1991. "Environmental Destruction in the Brazilian Amazon." In D. Goodman and A. Hall, eds., The Future of Amazonia: Destruction or Sustainable Development, pp. 179–225. New York: St. Martin's Press.

Feder, Ernest. 1971. The Rape of the Peasantry: Latin America's Landholding System. Garden City, N.Y.: Anchor Books.

Federacion de Centros Shuar. 1976. Solucion Original a un Problema Actual. Sucua.

Ferdon, Edwin N. 1950. Studies in Ecuadorian Geography. Monographs of the School of American Research, #15. Santa Fe, N.M.

Fifer, J. Valerie. 1982. "The Search for a Series of Small Successes: Frontiers of Settlement in Eastern Bolivia." Journal of Latin American Studies 14(2):407–432.

Findley, Sally E. 1988. "Colonist Constraints, Strategies, and Mobility: Recent Trends in Latin American Frontier Zones." In A. S. Oberai, ed., Land Settlement Policies and Population Redistribution in Developing Countries, pp. 271–316. New York: Praeger.

Foresta, Ronald A. 1991. Amazon Conservation in the Age of Development. Gainesville: University of Florida Press.

Foweraker, Joe. 1981. The Struggle for Land: A Political Economy of the Amazon Frontier in Brazil from 1930 to the Present Day. Cambridge, Eng.: Cambridge University Press.

Fundacion Natura. 1988. "Development Policy Issues for Ecuador's Amazonia." Paper prepared for the World Bank. Quito.

Garcia, Lorenzo. 1985. Historia de las Misiones en la Amazonia Ecuatoriana. Quito: Ediciones Abya-Yala.

Geddes, W. R. 1976. Migrants of the Mountains: The Cultural Ecology of the Blue Miao (Hmong Njua) of Thailand. Oxford: Clarendon Press.

Geertz, C. 1963. Agricultural Involution: Processes of Ecological Change in Indonesia. Berkeley: University of California Press.

Giddens, Anthony. 1984. The Constitution of Society: Outline of the Theory of Structuration. Berkeley: University of California Press.

Gildesgame, Myron L. 1989. "Conserving Tropical Forest Resources in the National Park System of Costa Rica." Ph.D. dissertation, Clark University.

Gill, Leslie. 1987. Peasants, Entrepeneurs, and Social Change: Frontier Development in Lowland Bolivia. Boulder and London: Westview Press.

Gladhart, P. 1972. "Capital Formation on the Ecuadorian Frontier: A Study of Human Investment and Modernization in the Riobambenos Cooperative." Cornell Agricultural Economics Paper #72-5, Ithaca.

Gonzalez, A. and J. Ortiz de Villalba. 1976. "Biografia de una Colonizacion: Km. 7-80 Lago Agrio—Coca." Centro de Investigaciones de la Amazonia Ecuatoriana (CICAME). Quito.

Gonzalez, S. R. and R. Morocho. 1979. "La Adaptabilidad de los Colonos de San Jose de Morona." Thesis, Facultad de Trabajo Social, Universidad Estatal de Cuenca, Cuenca.

Grindle, Merilee. 1986. State and Countryside: Development Policy and Agrarian Politics in Latin America. Baltimore: Johns Hopkins University Press.

Grubb, P. J. and T. Whitmore. 1966. "The Climate and Its Effects on the Distribution and Physiognomy of the Forests." Journal of Ecology 54(2):303–333.

Grubb, P. J., J. Lloyd, T. Pennington, and T. Whitmore. 1963. "A Comparison of Montane and Lowland Rain Forest in Ecuador." Journal of Ecology 51(3):567–601.

Guerrero, F. 1987. "Problemas Ecologicos y Sociales Relacionados con el Cultivo de Palma Africana: El Caso de Palmoriente." In F. Larrea, ed., Amazonia: Presente y . . . . ?, pp. 219–266. Quito: Ediciones Abya-Yala.

Guppy, Nicholas. 1984. "Tropical Deforestation: A Global View." Foreign Affairs 62:928–965.

Hall, Anthony. 1989. Developing Amazonia: Deforestation and Social Conflict in Brazil's Carajas Programme. Manchester: Manchester University Press.

Hamilton, C. S. 1984. Deforestation in Uganda. Nairobi: Oxford University Press.

Hamilton, L. S. 1987. "What Are the Impacts of Himalayan Deforestation on the Ganges-Brahmaputra Lowlands and Delta?: Assumptions and Facts." Mountain Research and Development 7(3):256–263.

Hardjono, Joan. 1986. "Spontaneous Rural Settlement in Indonesia." In United Nations Center for Human Settlements, Spontaneous Settlement Formation in Rural Areas. Vol. 2: Case Studies, pp. 50–69. Nairobi.

Harner, M. J. 1973. The Jivaro: People of the Sacred Waterfall. New York: Anchor Books.

Hartshorn, G. 1989. "Sustained Yield Management of Natural Forests: The Palcazu Production Forest." In J. Browder, ed., Fragile Lands of Latin America, pp. 130–138. Boulder: Westview Press.

Hazlewood, Peter T. 1986. "An Economic Perspective on Tropical Deforestation." Tulane Studies in Zoology and Botany 26:77–88.

Hecht, Susanna. 1985. "Environment, Development, and Politics: Capital Accumulation and the Livestock Sector in Eastern Amazonia." World Development 13(6):663–684.

——. 1989. "The Sacred Cow in the Green Hell: Livestock and Forest Conversion in the Brazilian Amazon." The Ecologist 19(6):229–234.

Hecht, S. and A. Cockburn. 1990. The Fate of the Forest. London: Verso.

Hegen, E. E. 1966. Highways Into the Upper Amazon Basin: Pioneer Lands in Southern Colombia, Ecuador, and Northern Peru. Gainesville: University of Florida Press.

Hennessy, Alistair. 1978. The Frontier in Latin American History. London: Edward Arnold.

Hendricks, Janet W. 1986. "Images of Tradition: Ideological Transformations Among the Shuar." Ph.D. dissertation, University of Texas.

Hermessen, J. L. 1917. "A Journey on the Rio Zamora, Ecuador." Geographical Review 4:434–449.

Hicks, J., H. Daly, S. Davis, and M. Lourdes de Frietas. 1990. "Ecuador's Amazon Region: Development Issues and Options." World Bank Discussion Paper #75, Washington, D.C.

Hill, Polly. 1963. The Migrant Cocoa Farmers of Southern Ghana. Cambridge, Eng.: Cambridge University Press.

Hiroaka, Mario. 1982. "The Development of Amazonia." Geographical Review 72(1):94–99.

Hiroaka, Mario and S. Yamamoto. 1980. "Agricultural Development in the Upper Amazon of Ecuador." Geographical Review 70(4):423–446.

Holdridge, L. R. 1967. Life Zone Ecology. Rev. ed. San Jose, Costa Rica: Tropical Science Center.

Hudson, J. C. 1969. " A Location Theory for Rural Settlement." Annals of the Association of American Geographers 59:365–381.

Hurst, Philip. 1990. Rainforest Politics: Ecological Destruction in South-East Asia. London: Zed Books.

Inman, Katherine. 1991. "Population, Development, Debt, and the Global Decline of Forests." M.A. thesis, Department of Sociology, University of Georgia.

Jacobs, Marius. 1988. The Tropical Rain Forest: A First Encounter. Berlin: Springer-Verlag.

James, Preston. 1959. Latin America. 3d ed. New York: Odyssey Press.

James, William E. 1983. "Settler Selection and Land Settlement Alternatives: New Evidence from the Philippines." Economic Development and Cultural Change 31(3):571–586, 1983.

Jaramillo Alvarado, P. 1936. Tierras del Oriente: Caminos, Ferrocarriles, Adminstracion, Riqueza Aurifera. Quito: Imprenta y Encuadernacion Nacionales.

——. 1964. Las Provincas Orientales del Ecuador: Examen Historico-Adminstrativo. Quito: Editorial Casa de la Cultura.

Jessup, Tim. 1989. "Forest Exploitation by Shifting Cultivators in Borneo (Kalimantan)." Manuscript. Rutgers University.

Johnson, B. and Wm. Clark. 1982. Redesigning Rural Development: A Strategic Perspective. Baltimore: Johns Hopkins Unversity Press.

Jolly, L. G. 1989. "The Conversion of Rain Forests to Pastures in Panama." In D. Schumann and W. Partridge, eds., The Human Ecology of Tropical Land Settlement in Latin America, pp. 86–130. Boulder: Westview Press.

Jones, Jeffrey. 1989. "Human Settlement of Tropical Colonization in Central America." In D. Schumann and W. Partridge, eds., The Human Ecology of Tropical Land Settlement in Latin America, pp. 43–85. Boulder: Westview Press.

Jordan, Carl F. 1987. Amazonian Rain Forests: Ecosystem Disturbance and Recovery. New York: Springer-Verlag.

Junta Nacional de Planificacion y Coordinacion Economica. 1965–1982, Anuario de Estadisticas Vitales. Quito.

Karsten, Rafael. 1935. The Head-Hunters of Western Amazonas: The Life and Culture of the Jibaro Indians of Eastern Ecuador and Peru. Societas Scientarium Fennico. Commentaciones Humanarium Litterarum, VI, Helsinki.

Kartawinata, K. and A. P. Vayda. 1984. "Forest Conversion in East Kalimantan, Indonesia: The Activities and Impact of Timber Companies, Shifting Cultivators, Migrant Pepper Farmers, and Others." In F. DiCastri, F. Baker, and M. Hadley, eds., Ecology in Practice, pp. 98–126. Dublin: Tycooly International Publishing.

Keyes, Charles. 1976. "In Search of Land: Village Formation in the Central Chi River Valley, Northeastern Thailand." Contributions to Asian Studies 9:45–65.

Kroeber, Alfred. 1948. Anthropology. New York: Harcourt, Brace.

Kunstadter, P. 1988. "Hill People of Northern Thailand." In J. Denslow and C. Padoch, eds., People of the Rain Forest, pp. 93–110. Berkeley: University of California Press.

Kunstadter, P., E. Chapman, and S. Sabhasri. 1978. Farmers in the Forest: Economic Development and Marginal Agriculture in Northern Thailand. Honolulu: University of Hawaii Press.

Lanly, J. P. 1983. "Assessment of the Forest Resources of the Tropics." Forestry Abstracts 44(3):287–313.

Lanly, J. P. and J. Clement. 1979. Present and Future Forest and Plantation Areas in the Tropics. Rome: FAO.

Lawson, Victoria. 1988. "Government Policy Biases and Ecuadorian Agricultural Change." Annals of the Association of American Geographers 78(3):433–452.

Ledec, George. 1985. "The Political Economy of Tropical Deforestation." In H. Jeffrey Leonard, ed., Divesting Nature's Capital: The Political Economy of Environmental Abuse in the Third World, pp. 179–226. New York: Holmes and Meier.

LeGrand, Catherine. 1984. "Colombian Transformations: Peasants and Wage Labourers in the Santa Marta Banana Zone." Journal of Peasant Studies 11(4):178–200.

——. 1986. Frontier Expansion and Peasant Protest in Colombia, 1830-1936. Albuquerque: University of New Mexico Press.

Lele, Ume and Steven Stone. 1989. "Population Pressure, the Environment, and Agricultural Intensification: Variations on the Boserup Hypothesis." MADIA Discussion Paper #4, World Bank, Washington, D.C.

Lewis, E. G. and H. T. Lewis. 1971. Ilocano Rice Farmers: A Comparative Study of Two Philippine Barrios. Honolulu: University of Hawaii Press.

Lisansky, J., 1990, Migrants to Amazonia: Spontaneous Colonization in the Brazilian Frontier. Boulder: Westview Press.

Logan, John R. and Harvey Molotch. 1987. Urban Fortunes: The Political Economy of Place. Berkeley: University of California Press.

Long, Norton. 1958. "The Local Community as an Ecology of Games." American Journal of Sociology 61(2):251–261.

MacDonald, Theodore. 1984. De Cazadores a Ganaderos. Quito: Ediciones Abya-Yala.

MAG (Ministerio de Agricultura y Ganaderia). 1977. La Colonizacion de la Region Amazonica Ecuatoriana: Obra Nacional. Quito.

MAG—Pronareg, 1980a. "Carta Pedo–Geomorfologica, Provincia de Morona Santiago, Parte Norte, Informe Provisional." Quito.

——. 1980b. "Las Zonas Socio-economicas Actualement Homogeneas de la Region Amazonica Ecuatoriana." Diagnostico Socio-economico del Medio Rural Ecuatoriano. Quito.

Maher, Dennis T. 1989. Government Policies and Deforestation in Brazil's Amazon Region. Washington, D.C.: World Bank.

Malingreau, Jean-Paul and C. J. Tucker. 1988. "Large-Scale Deforestation in the Southeastern Amazon Basin in Brazil." Ambio 17:49–55.

Margolis, Maxine. 1973. The Moving Frontier: Social and Economic Change in a Southern Brazilian Community. Gainesville: University of Florida Press.

Marsh, Robin R. 1983. Development Strategies in Rural Colombia: The Case of Caqueta. Latin American Center Publications, University of California at Los Angeles.

Martine, George. 1980. "Recent Colonization Experiences in Brazil: Expectations Versus Reality." In F. Barbira-Scazzicchio, ed., Land, People, and Planning in Contemporary Amazonia, pp. 80–94. Cambridge: Center of Latin American Studies.

Martz, John P. 1987. Politics and Petroleum in Ecuador. New Brunswick: Transaction Books.

Maxwell, Simon. 1980. "The Barbecho Crisis in Colonies of Santa Cruz, Bolivia." In F. Barbira-Scazziocchio, ed., Land, People, and Planning in Contemporary Amazonia, pp. 162–170. Cambridge: Center of Latin American Studies.

Migdal, J. S. 1988. Strong Societies and Weak States: State-Society Relations and State Capabilities in the Third World. Princeton: Princeton University Press.

Molotch, Harvey. 1976. "The City as a Growth Machine: Toward a Political Economy of Place." American Journal of Sociology 82(2):309–332.

Monbeig, P. 1966. "Les franges pionnieres." Geografie Generale, pp. 974–1006. Paris: Gallimard.

Moran, Emilio. 1987. "Monitoring Fertility Degradation of Agricultural Lands in the Lowland Tropics." In P. Little and M. Horowitz, eds., Lands at Risk in the Third World: Local Level Perspectives, pp. 69–91. Boulder: Westview Press.

——. 1989. "Adaptation and Maladaptation in Newly Settled Areas." In D. Schumann and W. Partridge, eds., The Human Ecology of Tropical Land Settlement in Latin America, pp. 20–39. Boulder: Westview Press.

Muller, K. D. 1974. Pioneer Settlement in South Brazil: The Case of Toledo, Parana. The Hague: Martinus Nijhoff.

Myers, Norman. 1984. The Primary Source: Tropical Forests and Our Future. New York: Norton.

Nair, P. K. R. 1990. "The Prospects for Agroforestry in the Tropics." World Bank Technical Paper #121, Washington, D.C.

Nations, J. D. and F. C. Hinojosa. 1989. "Cuyabeno Wildlife Production Reserve." In J. Browder, ed., Fragile Lands of Latin America, pp. 139–149. Boulder: Westview Press.

Navas, B. G. and F. Lara. 1978. "Organizacion Campesina: El Caso de Dos Organizaciones de Colonizacion en la Region Oriental del Ecuador." Desarrollo Rural en las Americas 10(1):47–65.

Nelson, Michael. 1973. The Development of Tropical Lands: Policy Issues in Latin America. Baltimore: Johns Hopkins University Press.

——. 1986. "Land Settlement in the Humid Tropics: Lessons from Experience in the Humid Tropics." In United Nations Center for Human Settlements. Spontaneous Settlement Formation in Rural Areas. Vol. 2: Case Studies, pp. 100–120. Nairobi.

Nicolai, H. and G. Lasserre. 1981. "Les Systemes de Cultures Traditionnels et les Phenomenes Pionniers en Afrique Tropicale." Travaux et Memoires de l'Institut des Hautes Etudes de l'Amerique Latine 34:95–115.

Noboa, Fernando J. 1990. Sancho Hacho: Origenes de la Formacion Mestiza Ecuatoriana. Quito: Ediciones Abya-Yala.

Ochoa, C. and L. Sierra. 1976. Un Comunidad Shuar en Proceso de Cambio. Sucua: Mundo Shuar.

Office of Technology Assessment. 1984. "Technologies to Sustain Tropical Forest Resources." Washington, D.C.: U.S. Congress.

Ortiz, Sutti. 1984. "Colonization in the Colombian Amazon." In Marianne Schmink and Charles Wood, eds., Frontier Expansion in Amazonia, pp. 204–230. Gainesville: University of Florida Press.

Padoch, Christine and W. de Jong. 1989. "Production and Profit in Agroforestry: An Example from the Peruvian Amazon." In J. Browder, ed., Fragile Lands of Latin America, pp. 102–113. Boulder: Westview Press.

Painter, M. and W. Partridge. 1989. "Lowland Settlement in San Julian, Bolivia: Project Success and Regional Underdevelopment." In D. Schumann and W. Partridge, eds., The Human Ecology of Tropical Land Settlement in Latin America, pp. 340–377. Boulder: Westview Press.

Parsons, J. J. 1976. "Forest to Pasture: Development or Destruction." Revista de Biologica Tropical 24(1):121–138.

Peluso, Nancy L. 1990. "A History of State Forest Management in Java." In Mark Poffenberger, ed., Keepers of the Forest: Land Management Issues in Southeast Asia, pp. 27–55. Hartford, Conn.: Kumarian Press.

Pelzer, Karl. 1945. Pioneer Settlement in the Asiatic Tropics: Studies in Land Utilization and Agricultural Colonization in Southeast Asia. New York: American Geographical Society.

——. 1978. "Swidden Cultivation in Southeast Asia: Historical, Ecological, and Economic Perspectives." In P. Kunstadter, E. Chapman, and S. Sabhasri, eds., Economic Development and Marginal Agriculture in Northern Thailand, pp. 271–288. Honolulu: University of Hawaii Press.

Perez Guerrero, Edmundo. 1954. Colonizacion e Immigracion en el Ecuador. Quito: Editorial Casa de la Cultura Ecuatoriana.

Platt, Ray R. 1932. "Opportunities for Agricultural Colonization in the Eastern Border Valleys of the Andes." In W. Joerg, ed., Pioneer Settlement: Comparative Studies by Twenty-Six Authors. Special Publication no. 14. New York: American Geographical Society.

Plumwood, Val and R. Routley. 1982. "World Rainforest Destruction: The Social Factors." The Ecologist 12(1):4–22.

Poffenberger, Mark. 1990. Keepers of the Forest: Land Management Alternatives in Southeast Asia. Hartford, Conn.: Kumarian Press.

Popkin, S. L. 1979. The Rational Peasant: The Political Economy of Rural Society in Vietnam. Berkeley: University of California Press.

Prance, G. 1973. "Phytogeographic Support for the Theory of Pleistoscene Forest Refuges Based on Evidence from Distribution Patterns in Caryocaraceae, Chrysobalanaceae, Dichapetalaceae, and Lecythidaccae." Acta Amazonica 3(3):1–26.

——. 1976. The Phytogeographic Subdivisions of Amazonia and Their Consequences for the Selection of Biological Reserves. Bronx, N.Y.: New York Botanical Garden.

Price, Marie. 1990. "Hands for the Coffee: Migration into Western Venezuela's Coffee Lands, 1870–1930." Paper presented at the Association of American Geographers Meetings. Toronto, Canada.

Price, Turner and Lana Hall. 1983. "Agricultural Development in the Mexican Tropics: Alternatives for the Selva Lacandona Region of Chiapas." Cornell Agricultural Economics Study #83-4.

Protang, Kaman and David E. Thomas. 1990. "Evolving Management Systems in Thailand." In M. Poffenberger, ed., Keepers of the Forest. pp. 167–186. Hartford, Conn.: Kumarian Press.

Raison, Jean-Pierre. 1968. "La Colonizacion des Terres Nueves Intertropicales." Etudes Rurales 31:5–112..

——. 1981. "La Colonizacion des Terres Nueves en Afrique Tropicale." Travaux et Memoires de l'Institut des Hautes Etudes de l'Amerique Latine 34:59–75.

Ramos, Mario. 1988. "The Conservation of Biodiversity in Latin America: A Perspective." In E. O. Wilson, ed., Biodiversity, pp. 428–436. Washington, D.C.: National Academy Press.

Redclift, Michael. 1987. Sustainable Development: Exploring the Contradictions. London and New York: Metheun.

——. 1989. "The Environmental Consequences of Latin America's Agricultural Development: Some Thoughts on the Brundtland Commission Report." World Development 17(3):365–377.

Repetto, Robert. 1989. The Forest for the Trees?: Government Policies and the Misuse of Forest Resources. Washington, D.C.: World Resources Institute.

Repetto, Robert and Malcolm Gillis, eds. 1988. Public Policies and the Misuse of Forest Resources. Cambridge: Cambridge University Press.

Reyes, O. and F. Teran. 1939. Historia y Geografia del Oriente Ecuatoriano. Quito: Talleres Graficos de Educacion.

Richards, J. F. and R. P. Tucker, eds. 1988. World Deforestation in the Twentieth Century. Durham: Duke University Press.

Richards, Paul F. 1952. The Tropical Rain Forest. Cambridge, Eng.: Cambridge University Press.

Rhoades, R. E. and Pedro Bidegaray. 1987. The Farmers of Yurimaguas: Land Use and Cropping Strategies in the Peruvian Jungle. Lima: International Potato Center.

Roberts, Ralph L. 1975. "Migration and Colonization in the Colombian Amazon: Agrarian Reform or Neo-latifundismo." Ph.D. dissertation, Syracuse University.

Rodriguez, Linda A. 1985. The Search for Public Policy: Regional Politics and Government Finances in Ecuador, 1830-1940. Berkeley: University of California Press.

Rudel, Thomas K. 1983. "Roads, Speculators, and Colonization in the Ecuadorian Amazon." Human Ecology 11:385–403.

——. 1989a. "Population, Development, and Tropical Deforestation: A Cross-National Study." Rural Sociology 54(3):327–338.

——. 1989b. "Resource Partitioning and Regional Development Strategies in the Ecuadorian Amazon." Geojournal 19(4):437–446.

Rudel, T. and Sam Richards. 1990. "Urbanization, Roads, and Rural Population Change in the Ecuadorian Andes." Studies in Comparative International Development 25(3):73–89.

Salazar, Ernesto. 1981. "The Federacion Shuar and the Colonization Frontier." In N. Whitten, ed., Cultural Transformations and Ethnicity in Modern Ecuador, pp. 589–613. Urbana: University of Illinois Press.
——. 1986. Pioneros de la Selva: Los Colonos del Proyecto Upano-Palora. Quito: Ediciones Abya-Yala.
Sandoval, F. 1987. "Petroleo y medio oriente en la Amazonia Ecuatoriana." In F. Larrea, ed., Amazonia: Presente y ...? pp. 174–196. Quito: Ediciones Abya-Yala.
Santana, Roberto. 1989. "Contraintes Agraires et Production Alimentaire en une Zone de Colonization (Sud-Est Equatorien)." In Crise Agricole et Crise Alimentaire dans les Pays Tropicaux, pp. 283–293. Paris: Comite Nacional du Geographie.
Sarmiento Chia and Marta Cecilia. 1985. "Meteti: Una Comunidad que Abre la Selva del Darien." In Stanley Heckadon Moreno and Jaime Espinosa Gonzalez, eds., Agonia de la Naturaleza, pp. 63–84. Instituto de Investigaciones Agropecuario de Panama.
Sawyer, Donald R. 1984. "Frontier Expansion and Retraction in Brazil." In Marianne Schmink and Charles Wood, eds., Frontier Expansion in Amazonia, pp. 180–204. Gainesville: University of Florida Press.
——. 1990. "The Future of Deforestation in Amazonia: A Socioeconomic and Political Analysis." In A. B. Anderson, ed., Alternatives to Deforestation, pp. 264–274. New York: Columbia University Press.
Schmink, Marianne and C. Wood, eds. 1984. Frontier Expansion in Amazonia. Gainesville: University of Florida Press.
——. 1987. "The Political Ecology of Amazonia." In P. D. Little and M. Horowitz, eds., Lands at Risk in the Third World: Local Level Perspectives, pp. 38–57. Boulder: Westview Press.
Scholz, Ulrich. 1986. "Spontaneous Rural Settlements and Deforestation in Southeast Asia." In United Nations Center for Human Settlements, Spontaneous Settlement Formation in Rural Regions. Vol. 2; Case Studies, pp. 13–25. Nairobi.
Schumann, Debra and William Partridge, eds. 1989. The Human Ecology of Tropical Land Settlement in Latin America. Boulder and London: Westview Press.
Schwartzman, S. 1989. "Extractive Reserves: The Rubber Tappers' Strategy for Sustainable Use of the Amazon Rainforest." In J. Browder, ed., Fragile Lands of Latin America, pp. 150–164. Boulder: Westview Press.
Scudder, Thayer. 1981. The Development Potential of New Lands Settlement in the Tropics and Subtropics: A Global State of the Art Evaluation with Specific Emphasis on Policy Implications. Binghampton: Institute for Development Anthropology.
Sedjo R. and M. Clawson. 1983. "How Serious Is Tropical Deforestation?" Journal of Forestry 81(12):792–794.

Sewastynowisc, James. 1986. "Two-Step Migration and Upward Mobility on the Frontier: The Safety Valve Effect in Pejibaze, Costa Rica." Economic Development and Cultural Change 34(4):730–753.

Seymour-Smith, Charlotte. 1988. Shiwiar: Identidad Etnica y Cambio en el Rio Corrientes. Quito: Ediciones Abya-Yala.

Shane, Douglas R. 1986. Hoofprints in the Forest: Cattle Ranching and the Destruction of Latin America's Tropical Forests. Philadelphia: Institute for the Study of Human Issues.

Shoemaker, Robin. 1981. The Peasants of El Dorado: Conflict and Contradiction in a Peruvian Frontier Settlement. Ithaca and London: Cornell University Press.

Shresta, Nanda. 1989. "Frontier Settlement and Landlessness Among Hill Migrants in Nepal's Terai." Annals of the Association of American Geographers 79(3):370–389.

Simkins, Paul D. and Frederick Wernstedt. 1971. Philippine Migration: The Settlement of the Digos-Padada Valley, Davao Province. Monograph Series #16, Yale University Southeast Asia Studies, New Haven.

Simmel, G. 1955. Conflict and the Web of Group Affliations. New York: Free Press.

Smith, Nigel. 1981. "Colonization Lessons from a Tropical Forest." Science 214:755–761.

——. 1982. Rainforest Corridors: The Transamazon Colonization Schems. Berkeley: University of California Press.

Southgate, D. 1990. "The Causes of Land Degradation Along Spontaneously Expanding Agricultural Frontiers in the Third World." Land Economics 66(1):93–101.

Spears, J. 1988. "Preserving Biological Diversity in the Tropical Forests of the Asian Region." In E. O. Wilson, ed., Biodiversity, pp. 393–402. Washington, D.C.: National Academy Press.

Spruce, Richard. 1908. Notes of a Botanist on the Amazon and the Andes. 2 vols. London: Macmillan.

Stearman, Allyn M. 1985. Camba and Kolla: Migration and Development in Santa Cruz, Bolivia. Orlando: University of Central Florida Press.

Stonich, Susan. 1989. "Social Processes and Environmental Destruction: A Central American Case Study." Population and Development Review 15(2):269–296.

Thoumi, Francisco E. 1990. "The Hidden Logic of Irrational Economic Policies in Ecuador." Journal of Interamerican Studies and World Affairs 32(2):43–68.

von Thunen, J. H. 1851. Recherches sur l'influence que le prix de grains, la richesse du sol, et les impots exercent sur les systeme de culture. M. Laverriere, trans. Paris: Guillaumin.

Tschopp, H. J. 1953. "Oil Exploration in the Oriente of Ecuador." Bulletin of the American Association of Petroluem Geologists 37(10):2303–2347.

Tucker, Richard P. and J. F. Richards, eds. 1983. Global Deforestation and the Nineteenth-Century World Economy. Durham: Duke University Press.

Uhlig, Harold. 1984. Spontaneous and Planned Settlement in Southeast Asia. Hamburg: Institute of Asian Affairs. Giessener Geographische Schriften, #58.

UNOTC (United Nations Office of Technical Cooperation). 1974. Evaluacion del Sistema de Planificacion Regional, Quito: Junta Nacional de Planificacion.

Utomo, K. 1967. "Villages of Unplanned Settlers in the Subdistrict of Kaliredjo, Central Lampung." In R. Koentjaraningrat, ed., Villages in Indonesia, pp. 281–298. Ithaca: Cornell University Press.

Vayda, A. P. 1983. "Progressive Contextualization: Methods for Research in Human Ecology." Human Ecology 11(3):265–281.

Vayda, A. P. and A. Sahur. 1985. "Forest Clearing and Pepper Farming by Bugis Migrants in East Kalimantan: Antecedents and Impact." Indonesia 39:93–110.

Veeck, Gregory and C. W. Pannell. 1989. "Rural Economic Restructuring and Farm Household Income in Jiangsu, People's Republic of China." Annals of the Association of American Geographer. 79(2):275–292.

Vining, Daniel R. 1986. "Population Redistribution Towards Core Areas of Less Developed countries, 1950–1980." International Regional Science Review 10(1):1–45.

Walker, Richard A. and M. K. Heiman. 1981. "Quiet Revolution for Whom?" Annals of the Association of American Geographers 71(1):67–83.

Wallace, Alfred Russel. 1969 (1853). A Narrative of Travels on the Amazon and Rio Negro. New York: Haskell House.

Watters, R. F. 1971. "Shifting Cultivation in Latin America." Food and Agriculture Organization (F.A.O.) Forestry Development Paper #17. Rome.

Weil, Connie. 1989. "Differential Economic Success Among Spontaneous Agricultural Colonists in the Chapare, Bolivia." In D. Schumann and W. Partridge, eds., The Human Ecology of Tropical Land Settlement in Latin America, pp. 264–297. Boulder: Westview Press.

Weil, Connie and Jim Weil. 1983. "Government, Campesinos, and Business in the Bolivian Chapare: A Case Study of Amazonian Occupation." Interamerican Economic Affairs 36:29–62.

Weil, J. 1989. "Cooperative Labor as an Adaptive Strategy Among Homesteaders in a Tropical Colonization Zone: Chapare, Bolivia." In D. Schumann and W. Partridge, eds., The Human Ecology of Tropical Land Settlement in Latin America, pp 298–339. Boulder: Westview Press.

Whitmore, T. C. 1984. Tropical Rain Forests of the Far East. 2d ed. Oxford: Clarendon Press.

Wilkie, D. 1988. "Hunters and Farmers of the African Forest." In J. Denslow and C. Padoch, eds., People of the Tropical Rain Forest, pp. 111–126. Berkeley: University of California Press.

Wilkie, J. and S. Haber. 1983. Statistical Abstract of Latin America, vol 22. Latin American Center Publications, University of California at Los Angeles.

Wilson, J., J. Hay, and M. Margolis. 1989. "The Bi-National Frontier in Eastern Paraguay." In D. Schumann and W. Partridge, eds., The Human Ecology of Tropical Land Settlement in Latin America, pp. 199–237. Boulder: Westview Press.

Winterbottom, R. 1990. Taking Stock: The Tropical Forest Action Plan After Five Years. Washington, D.C.: World Resources Institute.

Wolf, Eric. 1970. Peasant Wars in the Twentieth Century. New York: Harper & Row.

Wood, Charles. 1983. "Peasant and Capitalist Production in the Brazilian Amazon: A Conceptual Framework for the Study of Frontier Expansion." In Emilio F. Moran, ed., The Dilemma of Amazonian Development, pp. 259–277. Boulder and London: Westview Press.

Woodwell, G. M., R. Houghton, and W. Kovalick. 1987. "Deforestation in the Tropics: New Measurements in the Amazon Basin Using Landsat and NOAA Advanced Very High Resolution Radiometer Imagery." Journal of Geophysical Research 92:2157–2163.

World Bank. 1980. World Development Report. New York: Oxford University Press.

World Resources Institute. 1985. Tropical Forests: A Call to Action. Washington, D.C.

——. 1990. World Resources, 1990-91. New York: Oxford University Press.

Yaccino, Tom. 1988. " Colonization in Southeastern Ecuador: A Policy Evaluation of the Upano—Palora Semi-Directed Colonization Project." Senior thesis, Augustana College.

Index